Recent Advances and Innovative Developments in Transportation Geotechnics

Buddhima Indraratna · Cholachat Rujikiatkamjorn
Editors

Recent Advances and Innovative Developments in Transportation Geotechnics

Keynote Volume ICTG 2024

Springer

Editors
Buddhima Indraratna
Faculty of Engineering and Information Technology
Transport Research Centre
University of Technology Sydney
Sydney, NSW, Australia

Cholachat Rujikiatkamjorn
Faculty of Engineering and Information Technology
Transport Research Centre
University of Technology Sydney
Sydney, NSW, Australia

ISBN 978-981-97-8244-4 ISBN 978-981-97-8245-1 (eBook)
https://doi.org/10.1007/978-981-97-8245-1

This Springer imprint is published by the registered company Springer Nature Singapore Pte Ltd.
The registered company address is: 152 Beach Road, #21-01/04 Gateway East, Singapore 189721, Singapore

Preface

Transportation Geotechnics is indeed a crucial field that plays a pivotal role in the planning, design, construction, and maintenance of transportation infrastructure. The International Conference on Transportation Geotechnics (ICTG) has been instrumental in bringing together experts, researchers, and professionals to share knowledge, advancements, and best practices in this domain. The evolution of the conference, starting from its initiation in 2008 at the University of Nottingham, UK [1], under the auspices of ISSMGE's Technical Committee TC3 on Geotechnics of Pavements, highlights the growing importance of Transportation Geotechnics on a global scale. The subsequent renaming of the Technical Committee to TC202 on Transportation Geotechnics in 2009 further emphasises the specific focus on the geotechnical aspects related to transportation infrastructure.

The conferences held under TC202 have provided a platform for the exchange of ideas and the dissemination of research findings. The locations of these conferences, such as Sapporo, Japan (2nd ICTG 2011) [2], Guimarães, Portugal (3rd ICTG 2016) [3], and Chicago, USA (4th ICTG 2020) [4], indicate the international scope and significance of Transportation Geotechnics. These events serve as opportunities for professionals from different parts of the world to collaborate, discuss challenges, and propose innovative solutions for the development and maintenance of safe and sustainable transportation networks.

The continued organisation of such conferences reflects the commitment of the geotechnical engineering community to advancing the field of Transportation Geotechnics. The shared knowledge and experiences contribute to the improvement of design methodologies, construction techniques, and maintenance practices, ultimately enhancing the safety and efficiency of transportation infrastructures globally.

The 5th International Conference on Transportation Geotechnics (5th ICTG 2024) was held in Sydney, Australia, in November 2024. The ICTG series is run by the Technical Committee (TC202) on Transportation Geotechnics of International Society of Soil Mechanics and Geotechnical Engineering (ISSMGE), under the chairmanship of Prof. Tatsuya Ishikawa (2022–2026). The Conference was organised by the Transport Research Centre of the University of Technology Sydney (UTS-TRC),

TC202 chaired by Prof. Tatsuya Ishikawa (2022–2026) and the Australian Geomechanics Society (AGS). The Conference was chaired by Distinguished Professor Buddhima Indraratna, Director UTS-TRC. Additional support was provided by the Office of Chief Scientist and Engineer, NSW, and Transport for New South Wales (TfNSW).

The 5th ICTG has also received further support through the memberships of the International Geosynthetics Society (IGS), Transport Research Board (TRB), Geo-Institute and USUCGER, Japanese Geotechnical Society (JGS), and the Southeast Asian Geotechnical Society (SEAGS). It involves efforts from three Technical Committees of the ISSMGE (TC202-Transportation Geotechnics, TC211-Ground Improvement and TC307-Sustainability).

The Conference aimed at promoting concepts and applications predominantly in the fields of transport infrastructure design, ground improvement, sustainability and circular economy perspectives, and natural hazards mitigation.

This book contains three keynote papers and 19 invited theme papers. Each submission was reviewed by at least one reviewer. Other keynotes, State-of-the-Art, State-of-Practice, and Heritage papers, were published in the Special Issue of the Transportation Geotechnics Journal (Guest Editors: Prof. Cholachat Rujikiatkamjorn, Prof. Tatsuya Ishikawa, Dr. Monica Prezzi and Prof. Mike Winter). The 5th Proctor Lecture paper by Prof Erol Tutumluer was published in the Proctor Series of the Transportation Geotechnics Journal.

In closing, we extend our heartfelt gratitude to all the contributors whose high-quality papers were reviewed by voluntary reviewers and overseen by the convenors of each chapter. The success of this book is indebted to the motivation and dedication of the authors, as well as the tireless efforts of the convenors and numerous members across various Committees, particularly the Local Organising Committee, Technical Committee, Reviewers, and International Advisory Committee.

We hope that you will find the contents of this Proceedings of valuable use in this respect.

Sydney, Australia

Buddhima Indraratna
Cholachat Rujikiatkamjorn

References

1. Ellis E, Yu H-S, McDowell G, Dawson AR, Thom N (eds) (2008) Advances in transportation geotechnics. Taylor and Francis Group, Boca Raton
2. Miura S, Ishikawa T, Yoshida N, Hisari Y, Abe N (eds) (2012) Advances in transportation geotechnics 2. Taylor and Francis Group, Boca Raton
3. Correia AG (ed) (2016) Advances in transportation geotechnics 3. Procedia Eng 143
4. Tutumluer E, Nazarian S, Al-Qadi I, Qamhia IIA (eds) (2022) Advances in transportation geotechnics 4. Springer Nature, Switzerland

Contents

About the Editors

Prof. Buddhima Indraratna is Distinguished Professor and Director at the Transport Research Centre, University of Technology Sydney (UTS). His pioneering work includes advancements in technical standards and practice guides, large-scale experimental modelling, computational design methods, and innovative design solutions for industry in the field of geotechnical and transportation engineering. For his significant contributions to Civil Engineering through infrastructure development, ground improvement, and transportation geotechnics, Prof Indraratna was bestowed the prestigious state honour of Member of Order of Australia on the Australia Day (26 January 2024) by the Governor General. His salient research outcomes have attracted many other national and international awards, also recognising his pioneering efforts in the use of waste materials for transport infrastructure. For his service and contributions to industry and community, including engagement in numerous infrastructure projects worldwide, he was awarded Honorary Life Membership of the Australian Geomechanics Society in 2023, International Civil Engineer Award by the Jose Entrecanales Ibarra Foundation (Madrid, Spain) in 2024, Sir John Holland Civil Engineer of the Year 2024 by the Institution of Engineers Australia, and numerous accolades from the International Society for Soil Mechanics and Geotechnical Engineering (ISSMGE), Institution of Civil Engineers (UK), International Association for Computer Methods and Advances in Geomechanics (IACMAG), and the Canadian Geotechnical Society. He is Fellow of the prestigious Australian Academy of Technological Sciences and Engineering (FTSE) and other professional organisations (FIEAust, FASCE, FGS) and Honorary Professor of several international institutions in Thailand, China, and India. He has written over 1000 publications including over 500 peer-reviewed journal articles and 13 books. He has supervised more than 80 Ph.D. students and mentored over 40 postdoctoral fellows through many research grants in the past 3 decades in Australia.

Prof. Cholachat Rujikiatkamjorn obtained his Ph.D. from the University of Wollongong, Australia, specialising on effective ground improvement solutions for low-lying floodplains and coastal terrains for designing robust transport infrastructure. Currently, he is Professor of Geotechnical Engineering and Program Leader of the UTS Transport Research Centre. He has been involved in real-life projects based on waste materials including reclamation works, highway maintenance and repairs, construction of access roads and railways using granular waste materials derived from local coal mining sites and steel manufacturing plants. He has published nearly 300 articles in international journals and conferences. Through numerous external grants over the years, he had supervised many Ph.D. students in the emerging fields of ground improvement and transport geotechnics, including the characterisation of waste materials such as coal wash, steel slag, shredded rubber, and crushed glass. For his outstanding contributions to the field of geotechnical engineering, he received the prestigious ISSMGE's Young Member Award in 2013, IACMAG's regional contribution medal in 2017, and Hugh Trollope Medal by the Australian Geomechanics Society (AGS) in 2012. He is Fellow of the Institution of Engineers Australia and currently holds Chair of AGS Sydney Chapter.

Bio-cementation for Ground Improvement with Special Reference to Transportation Infrastructure

Mohamed A. Shahin, Kehinde Lemboye, Liang Cheng, Hayder H. Abdullah, and Mohamed G. Arab

Abstract Unstable soils, characterised by their loose, soft, erodible and collapsible nature, are prevalent worldwide, presenting significant engineering challenges due to their low-bearing capacity and high compressibility. Such problematic soils pose obstacles to civil infrastructure developments such as building foundations, roads, railways, tunnels and retaining walls. The traditional approach to mitigating these obstacles involves ground improvement techniques that utilise soil stabilisation with chemical additives such as lime or Portland cement. While these techniques enhance soil strength, stiffness and load-bearing capacity, they also cause serious environmental concerns. In this Theme Lecture paper, "bio-cementation" is introduced as an innovative and eco-friendly technology for ground improvement, with special reference to transportation infrastructure. This ground-breaking technology utilises earth-based bacteria or microorganisms for natural biological soil stabilisation, paving the way for superior and eco-friendly ground improvement in future civil engineering infrastructure projects.

Keywords Ground improvement · Soil stabilisation · Bio-cementation · Crushed rock base · Road base material · Transportation infrastructure

M. A. Shahin (✉) · K. Lemboye · H. H. Abdullah
School of Civil and Mechanical Engineering, Curtin University, Perth, Australia
e-mail: m.shahin@curtin.edu.au

L. Cheng
School of Environment and Safety Engineering, Jiangsu University, Zhenjiang, China

M. G. Arab
Department of Civil and Environmental Engineering, College of Engineering, University of Sharjah, Sharjah, United Arab Emirates

B. Indraratna and C. Rujikiatkamjorn (eds.), *Recent Advances and Innovative Developments in Transportation Geotechnics*,
https://doi.org/10.1007/978-981-97-8245-1_1

1 Introduction

The stability of soil structure is fundamentally determined by the nature of the bonds that link the soil particles. Such bonds may be either inherent like the ionic, covalent and hydrogen bonds or naturally developed at the interfaces between soil particles through the precipitation of substances like calcite, silica, alumina and iron oxides [1]. Nonetheless, not all soils exhibit strong bonding, and certain soil types like loose, soft, erodible and collapsible soils, often exhibit problematic behaviour, particularly when exposed to water. This is attributed to the tendency of such problematic soils to expand or compress excessively, even under minimal loads. This problem presents significant challenges for geotechnical engineers and necessitates ground improvement before construction [2]. Soil improvement is typically achieved using chemical stabilisation through the introduction of a chemical binder to foster the artificial cementation bonds between soil particles. Ordinary Portland Cement (OPC) is the most commonly used soil stabilisation binder; however, the production of this chemical binder is responsible for approximately 6–7% of global artificial carbon dioxide emissions annually [3]. Other environmental challenges associated with the OPC industry include the sourcing of raw materials and over-reliance on coal quarries for energy. Consequently, in the context of global warming, the extensive use of OPC in soil stabilisation further exacerbates its negative environmental impact. Given the current concerns around global warming, the widespread use of OPC in large-scale operations like soil stabilisation has a further negative effect on the environment. This pressing issue has spurred the creation of alternative binders that not only have a reduced environmental impact but also maintain the effectiveness of soil stabilisation [4–6].

Bio-cementation has recently been identified as a promising technology for enhancing soil geomechanical properties. This innovative approach draws inspiration from natural biological processes, wherein living organisms utilise a mix of organic and inorganic compounds to produce robust and rigid materials between soil particles [7]. The implementation of bio-cementation can be achieved through two primary methods known as Microbial-Induced Carbonate Precipitation (MICP) and Enzyme-Induced Carbonate Precipitation (EICP). The latter, EICP, was conceived to address certain challenges posed by MICP, simplifying field implementation while also reducing treatment costs. The emergence of bio-cementation marks a significant advancement in ground improvement technology, offering a sustainable and environmentally friendly alternative to traditional soil stabilisation methods. While this green technology signifies a notable step towards a more sustainable future for the industry, it still requires further research and development to overcome a few existing challenges to fully commercialise its use and satisfy industrial needs.

This paper aims to provide an overview of bio-cementation as a soil stabilisation process, with a particular focus on transportation infrastructure. The paper explores the use of EICP as one of the effective bio-cementation methods and evaluates its geotechnical attributes when applied to the treatment of crushed road base material, based on preliminary experimental work. The paper also highlights the benefit of bio-cementation on the brink of revolutionising soil stabilisation practices, heralding an era of transformation for the ground improvement industry.

2 Bio-cementation Techniques

2.1 MICP Versus EICP

Bio-cementation is an ecological process that results in the deposition of calcium carbonate or calcite into soil voids to link soil particles together. As identified by Kavazanjian and Karatas [8], this process can be stimulated through various mechanisms, including urea hydrolysis, microbial denitrification and sulphate reduction. However, among available mechanisms, urea hydrolysis, due to its relative simplicity, has been the subject of extensive research and application. The implementation of urea hydrolysis in bio-cementation entails the utilisation of bacteria, such as *Sporosarcina pasteurii*, which act as a source of the urease enzyme. This particular process, known as Microbial-Induced Carbonate Precipitation (MICP), has been investigated for an array of soil improvement applications such as erosion resistance, slope stability, under-seepage of levees and enhancing bearing capacity of foundations [9, 10]. Nevertheless, the practical application of MICP faces some limitations due to various challenges. These encompass the requirement for multiple treatment cycles to attain satisfactory carbonate precipitation and the limitation of bacterial size, which hinders permeability in some soil types [11, 12]. Additionally, bacterial growth and enzyme production necessitate specific environmental conditions such as temperature, pH and oxygen availability, further complicating the application of MICP [13]. To overcome such challenges, an alternative method, Enzyme-Induced Carbonate Precipitation (EICP), has been proposed. EICP employs free urease enzymes derived from agricultural or bacterial sources [14], acting as a catalyst in hydrolysis. EICP obviates the need for nutrients to stimulate bacterial activity, thereby simplifying field application and decreasing treatment costs compared to MICP. Additionally, the absence of living organisms in EICP implies that it is not influenced by the cellular processes or metabolic rates specific to microbial species [15].

Comparisons between MICP and EICP at a microscale level can be challenging due to the differing behaviour of the free enzyme and living bacteria, influenced by factors such as soil type, cementing solution concentration and enzyme source. However, in terms of their performance and efficiency, a comparison can be deduced from standard shear and unconfined compressive strength testing. A study conducted by Nafisi et al. [14] demonstrated that EICP-treated soils required less calcium carbonate precipitation to achieve the same shear wave velocity as that obtained from MICP-treated soils. Even though MICP resulted in higher shear strength and larger dilative strain, it required a higher number of injections to reach the same shear wave velocity as EICP-treated soils. In the realm of MICP, various strategies have been proposed to stimulate higher carbonate precipitation. One such strategy involves the implementation of multiple cycles of two-phase injection, which first introduces a microbial culture and then a cementing solution [5, 16]. In the context of EICP, however, an augmented level of precipitated carbonate can be achieved following a single cycle of treatment. This is made possible through the one-phase-low-pH method proposed by Cui et al. [17], which involves injecting an EICP solution into the soil for a singular phase of sand treatment. The use of a high concentration of EICP solution has also been proposed as an effective method [18]. Nevertheless, when comparing the two methods, the use of microbially produced urease in MICP has demonstrated superior results over the free urease enzymes utilised in EICP. This superiority is notably observed in the enhancement of soil mechanical properties [19]. Despite this limitation, the logistical advantages of EICP are markedly superior to those of MICP. One of the significant contrasts between the two methods lies in the stringent temperature controls and handling measures required by MICP. EICP presents an advantage by allowing enzyme transportation in dry powder form compared to MICP which requires bacteria transportation in liquid or slurry form, the benefits of EICP in this regard become more pronounced. This feature not only simplifies the treatment process but also potentially leads to considerable cost savings. As this paper progresses, the focus will be redirected towards the application of EICP in soil stabilisation, with special reference to transportation infrastructure.

2.2 *EICP for Soil Stabilisation*

The EICP reaction in stabilised soil essentially depends on the hydrolysis of urea, which provokes the precipitation of calcium carbonate among soil particles, provided that there is an adequate presence of calcium ions. The following section provides an exploration of the precipitation mechanism and the corresponding affecting factors.

2.2.1 Precipitation Mechanism

The process of urease-aided carbonate mineralisation is primarily catalysed by the enzyme urease, which facilitates the hydrolysis of urea. This reaction results in the formation of ammonia and carbonate. Such substances, due to their instability, further degrade into a second molecule of ammonia and carbonic acid. In the presence of water, the resulting products transform into bicarbonate, ammonium and hydroxide ions, respectively. The ion production leads to an increase in pH, which subsequently promotes the formation of carbonate ions [20]. This entire procedure is encapsulated in Eq. (1). The pH elevation, combined with the presence of dissolved Ca^{2+}, fosters favourable conditions for ion merging, ultimately leading to the precipitation of calcium carbonate, as given below in Eqs. (2) and (3).

$$CO(NH_2)_{2(s)} + 2H_2O \xrightarrow[]{\text{Urease enzyme}} 2NH^+_{4(aq)} + CO^{2-}_{3(aq)} \quad (1)$$

$$CaCl_{2(s)} \xleftrightarrow{H_2O} Ca^{2+}_{(aq)} + 2Cl^-_{(aq)} \quad (2)$$

$$Ca^{2+}_{(aq)} + CO^{2-}_{3(aq)} \xleftrightarrow{\text{precipitation}} CaCO_{3(s)} \quad (3)$$

The urease employed by EICP can be sourced from either bacterial or agricultural origins. Although both sources are viable, agricultural sources are predominantly used in EICP applications. Various agricultural sources such as watermelon, soybean, pumpkin, winged bean and jack bean have been employed to derive urease [20]. However, the effectiveness of EICP is highly contingent on the source and activity of urease, which can vary significantly with environmental pH, temperature and concentration. Therefore, several researchers have introduced various additives to the EICP cementing solution to enhance the EICP performance. The selection of the urease source is a critical factor in the EICP process and continued research is necessary to identify cost-effective and efficient alternatives to pure urease [20].

2.2.2 Perception of Influencing Factors

The efficiency and effectiveness of the EICP process in precipitating calcium carbonate are influenced by various factors, including the concentration of chemical constituents, temperature and pH level. The concentration of chemical constituents, particularly urea, is vital in the EICP process. The urease enzyme catalyses the hydrolysis of urea, leading to the formation of calcium carbonate. Studies have shown that increasing the concentration of urea can lead to increased hydrolysis rates, up to a certain threshold. Beyond this threshold, the presence of inhibitors, particularly ammonium ions, can lead to a decrease in hydrolysis rates [21]. Therefore, optimising the concentration of urea is crucial for maximising the efficiency of the EICP. Temperature is another factor that can significantly affect the EICP process. According to Nemati and Voordouw [22], increasing the temperature from 20 to 50 °C enhances the production rate of calcium carbonate, which can be attributed to the increased activity of the urease enzyme at higher temperatures. Lower temperatures, however, can lead to a decrease in enzyme activity, thus reducing the efficiency of the EICP [19]. The pH level also plays a pivotal role in the EICP process. The hydrolysis of urea leads to an increase in the pH level, shifting the carbonate equilibrium and leading to the precipitation of calcium carbonate. A high pH level (> 9.0) is beneficial for increasing calcium carbonate saturation and driving the precipitation reaction towards completion [23]. Additionally, the activity of the urease enzyme is also dependent on the pH level, with optimal activity observed at a pH of 6.9 [24].

In addition to the abovementioned factors, the efficiency of the EICP process is significantly impacted by the physical characteristics inherent in the host soil. Among these, the shape of the soil particles is of significant relevance. Studies indicate that the performance of bio-cemented soils is contingent upon not only the soil relative density but also the shape and distribution of the soil particles [21, 25, 26]. Reports suggest that rounded particles are more conducive to carbonate precipitation adherence, thereby increasing the carbonate content. However, the effect of soil shape on the compressive strength of the soil remains a subject of ongoing debate [27]. Soil type is another influential factor. Most EICP studies have focused on clean sand, with fewer studies examining the application of EICP in silty and clayey soils. Chandra and Ravi [28] found that EICP was ineffective in treating silty soils due to low pH, which negatively affects urea hydrolysis and calcite precipitation. Conversely, Oliveira et al. [29] found that silty soil with a pH of 7.75 showed a significant improvement in the EICP precipitation ratio. The presence of nucleation sites for carbonate precipitation is also a key consideration. The nucleation sites aid in forming effective bridges between soil contact points, and their absence in the EICP can negatively impact the distribution of carbonate precipitation within soil pores [30]. Proposals have been made to use non-fat powder milk as an enzyme stabiliser and calcium ion binder, thereby acting as nucleation sites for carbonate precipitation. The morphology of carbonate crystals also plays a crucial role in the effectiveness of the EICP-treated soils. Carbonate crystals that form effective bridges between soil particles significantly enhance soil strength. The morphology and location of crystals can be assessed using SEM images [4]. Studies have shown that disturbance in calcite crystal formation can occur in the presence of inhibiting matter such as magnesium ions and organic compounds [21].

3 Utilisation of EICP in Transportation Infrastructure

The three main layers that make up a road pavement are the surface asphalt course, base course and subgrade or natural soil layer. Roads are typically built from the bottom up, with the subgrade at the bottom and the base material compacted on top of the subgrade before the surface course is installed. Within the context of transportation infrastructure, research on the use of EICP proved to be advantageous in road applications when assessing the quality of the subgrade layer. However, there is a noticeable lack of research focused on coarse aggregate, the principal material used for base and subbase layers. This gap of knowledge will be investigated later in this paper through some limited preliminary experimental work.

In terms of the application of EICP for soil stabilisation, four primary methods have been identified in the literature for the introduction of the EICP cementing solution. These include injection, surface percolation, spraying and mix-and-compact. To assess the viability of these methods within the transportation infrastructure sector, a brief evaluation of their practicality and efficiency for soil stabilisation is warranted. *The injection method* entails a direct infusion of the EICP solution into the soil. Hamdan [30] demonstrated the efficacy of this method on Ottawa (20–30) silica sand columns, with the recorded unconfined compressive strength (UCS) values ranging from 35 to 125 kPa. Kavazanjian et al. [31] also utilised this method as a grouting solution for soil nails. However, Neupane et al. [32] noted potential issues with uneven carbonate distribution and increased precipitation near the injection point, potentially reducing the porosity and effectiveness of the subsequent injections. *The surface percolation method* involves the application of the EICP solution on the soil's top surface. Hamdan [30] adopted this method for sand column treatment, achieving UCS values between 38 and 210 kPa. Multiple applications were required to enhance carbonate precipitation and compressive strength. Almajed et al. [21] proposed that this method promotes nucleation and growth between soil particles, offering superior binding. *The spraying method* is typically employed when the cementing solution must remain on the treated soil's surface. It is particularly useful for applications such as water erosion resistance and fugitive dust control. An example of the effectiveness of this method is demonstrated by Knorr [33] for dust control, resulting in soil surfaces that could withstand wind speeds up to 24.4 m/s. *The mix-and-compact method* encompasses the mechanical blending of the EICP solution with the soil, accompanied by compaction. Hamdan [30] demonstrated the efficacy of this method in treating sand specimens, by achieving UCS values from 392 to 529 kPa, markedly higher than those achieved using the injection method. In conclusion, the choice of an appropriate EICP application method for soil stabilisation is significantly influenced by factors such as soil grain size and pore size. Each technique, including injection, surface percolation, spraying and mix-and-compact, has unique applications.

The abovementioned EICP methods showcase a substantial potential for advancements in surficial soil stabilisation using EICP and hold particular significance for their application in transportation infrastructure, including flexible pavements. This could lead to the enhancement and prolongation of the structural integrity of flexible pavements through the stabilisation of the subgrade, subbase and base course using EICP. Austroads guidelines [34] stipulate that pavement materials, classified as lightly bound, should be stabilised with an OPC content of less than 3%, which yields UCS of 1–2 MPa. However, to achieve this level of strength, some challenging scenarios necessitate an OPC content of 3% or more. This substantial requirement to attain such high UCS values often triggers the development of shrinkage cracks during OPC hydration [35]. Such cracks potentially propagate to the asphalt top layer, creating conduits for moisture ingress, which can potentially lead to the deterioration of the entire pavement structure. Therefore, EICP could serve as a potent alternative to OPC, mitigating the risk of potential cracks within treated soil, which is essential for achieving high UCS values. Overall, implementing the EICP technique in the transportation infrastructure sector can yield several advantages over conventional methods. Notably, EICP holds the potential to significantly bolster environmental sustainability. In contrast to the widely used OPC, which is notorious for its high carbon dioxide emissions, the carbonate cementation facilitated by EICP presents a more environmentally friendly solution for bulk applications such as pavement construction [11]. Moreover, the EICP process is less energy-demanding than OPC, which further amplifies its environmental credentials. Despite these advantages, the application of EICP also presents certain challenges. Environmental concerns, for instance, arise from the production of ammonium chloride as a by-product of the

EICP process. This compound can be potentially harmful and may contaminate groundwater supplies [33, 36]. The decision to flush the ammonium from the soil may not be economically viable. Another challenge lies in achieving distribution and homogeneity in the EICP treatment. The water-based, low-viscosity nature of the EICP solution makes it difficult to control the treated soil mass. Possible solutions to this problem include increasing the viscosity of the treatment solution using biopolymers or copolymers to enhance its distribution along the soil matrix [37].

4 Stabilisation of Crushed Rock Base Using EICP—Experimental Investigation

Most of the existing research on the use of EICP for transportation infrastructure concentrates on sand, which represents the road subgrade layer. There is a conspicuous absence of studies on the use of EICP for other road layers, including the base or subbase layers. Therefore, the potential use of EICP for the treatment of crushed rock base constitutes the principal focus of this paper. This investigation also contributes to an emerging field of research, with unique novelty stemming from the synthesised recipe of EICP that incorporates the use of crude urease enzyme derived from soybeans. Based on some preliminary experimental work, the strength development of the treated crushed rock base under uniaxial loading conditions was examined using UCS tests. Additionally, the efficacy of the EICP application was assessed through the evaluation of calcium carbonate precipitation. Furthermore, an exploration was undertaken to discern the influence of soybean concentration on the geomechanical behaviour of the EICP-treated CBR material. It should be noted that research is ongoing by the authors for a full investigation of this promising stabilisation technology for road base materials including mitigation of tensile failure, durability, long-term performance and economic viability, and the results will be published in due time.

4.1 Material and Methodology

4.1.1 Road Base Material

This study utilises crushed rock base (CRB), a local road base course material available in Western Australia, as the road base aggregate. A preliminary evaluation of the CRB aggregate used was conducted through a particle size distribution test to ascertain its suitability for road construction. The particle size distribution test was executed following Australian Standard AS1289.3.6.1-2009 [38] and the ensuing grading was verified against the base course quality thresholds specified in the Main Road Western Australia MRWA [39]. The gradation of the CRB aggregate used

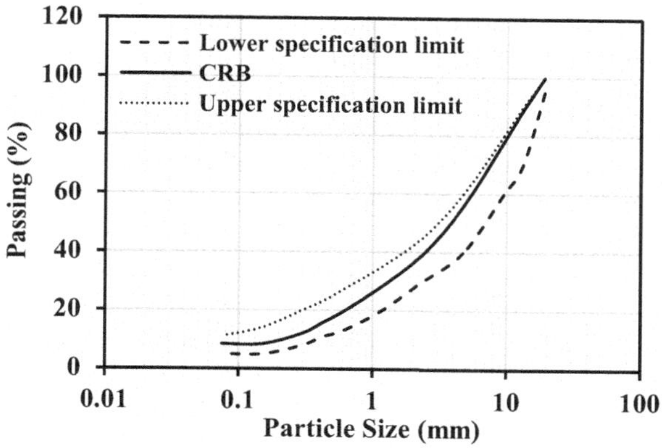

Fig. 1 Particle size distribution of crushed rock base (CRB) used in this study

varied from 0.150 to 13.2 mm, indicating a diverse range of particle sizes, and a graphical representation of the particle size distribution of the CRB used is provided in Fig. 1.

4.1.2 Sample Preparation

As mentioned earlier, the EICP reaction in soil stabilisation essentially depends on the hydrolysis of urea, which provokes the precipitation of calcium carbonate among soil particles, provided that there is an adequate presence of calcium ions. To prepare the EICP treatment, ingredients such as urease enzyme, urea, anhydrous calcium chloride and non-fat milk were utilised. The process was facilitated by the urease enzyme, which catalyses the conversion of urea into carbonate and ammonium ions. For this particular study, the chosen urease enzyme was derived from soybeans. This involved initially grinding 150 g of soybean into powder using an electric grain mill grinder. Following this, a specific weight of the pulverised soybean was combined with deionised water. Calcium sulphate was included in the mixture to assist in salting out any excess organic matter that could potentially disrupt the precipitation of calcium carbonate. After allowing the mixture to stand undisturbed for 30 min, it was thoroughly blended until there were no lumps, resulting in a homogenous mixture. This mixture was then placed into centrifuge tubes and subjected to centrifugation at 4000 RPM for 15 min at a temperature of 4 °C. The resulting supernatant was identified as the crude urease enzyme. The level of urease activity was ascertained by evaluating the rate of ureolysis in a solution composed of 5 mL of 3 M urea, 4 mL of deionised water and 1 mL of soybean crude enzyme. This measurement was conducted using the electrical conductivity method, as suggested by Whiffin [40]. The study took into consideration a soybean concentration of 75, 125 and 175 g/L, which correspond to urease activities of 8, 15 and 20 U/mL, respectively. The current study examined a modified EICP treatment process by dissolving 2 M of cementation solution of urea and calcium chloride in deionised water, followed by the addition of 12 g/L of non-fat or skim milk. The inclusion of milk not only stabilises the urease enzyme but also enhances the nucleation of the precipitated calcium carbonate [21]. The cementation solution (urea + calcium chloride) was then pre-mixed with the desired concentration of the soybean crude enzyme in equal proportions and introduced to the CRB samples. Upon mixing, the final concentration of the EICP cementation solution was 1 M, with the milk concentration standing at 6 g/L.

To produce the EICP-stabilised CRB samples, cylindrical moulds were fabricated from polyvinyl chloride (PVC) pipes. These moulds measured 105 mm in diameter and 115.5 mm in height following Australian Standard AS5101.4-2008 [41]. The process of preparing the EICP-treated CRB samples involved blending the CRB with water at optimum moisture content, followed by gentle compaction in pre-fabricated polyvinyl chloride moulds, resulting in a density of 1.67 g/cc and a porosity level of 34%. Subsequently, the EICP treatment solution was introduced, amounting to 1.12 times the pore volume from the top of the CRB samples, and permitted to seep down to the bottom. Multiple treatment cycles were undertaken, and seven cycles were found to reach the desired cementation level. The treatment involved alternating cycles of EICP solution (urease enzyme + urea + calcium chloride) and cementation solution (urea + calcium chloride). The EICP solution was applied in one cycle and an exclusive cementation solution in the next cycle and so on until the seven cycles were applied. Each treatment cycle was given a 24 h reaction period after application.

The effluent from each sample was collected after every treatment cycle to assess the chemical conversion efficiency (CCE) of the reaction. The Nessler reagent method was utilised to measure the ammonia concentration following the hydrolysis of urea by Whiffin [40]. Post-treatment, the samples were flushed with two pore volumes of distilled water to eliminate the ammonia by-product arising from the chemical reaction. Following this, the samples were removed and stored before testing.

Additional samples were also prepared and stabilised by adding 2 and 3% OPC to CRB. These samples were tested for comparison with the EICP-stabilised samples. The admixture was compacted in five layers to achieve a maximum dry density of 2.2 g/cm^3 and an optimum moisture content of 5.7%. The test samples were covered with cling film and allowed to cure for 7, 14 and 28 days before testing. After each curing period, the samples were soaked in water at a temperature of 25 °C for 4 h and allowed to drain for 15 min before conducting the UCS test.

4.1.3 Tests Conducted

A strain-controlled unconfined compressive strength (UCS) test was conducted to assess the stress–strain behaviour of the EICP-stabilised and OPC-stabilised CRB samples. The test was conducted following the Australian Standard AS5101.4-2008 [41] and was carried out at a 1.0 mm/min displacement rate. Triplicate chunks of EICP-stabilised CRB samples were randomly collected from the intact and failure

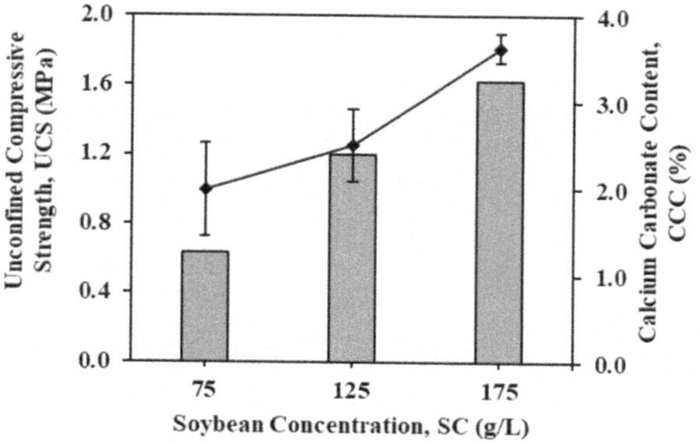

Fig. 2 Effect of soybean concentration on strength and calcium carbonate content of EICP-cemented CRB samples

zone after the UCS test to determine the amount of calcium carbonate content (CCC) precipitated. The CCC was obtained by measuring the volume of carbon dioxide gas that evolved with brisk effervescence following the reaction of the calcium carbonate present in the EICP-cemented CRB samples with concentrated hydrochloric acid.

5 Results and Discussion

5.1 Effect of Soybean Concentration

Figure 2 illustrates the impact of soybean concentration (SC) on both the unconfined compressive strength (UCS) and calcium carbonate content (CCC) of EICP-cemented CRB specimens, which underwent seven cycles of alternating treatments of EICP solution and cementation solution with a 1 M cementation solution concentration. As can be seen, a discernible linear correlation between the SC and both the UCS and CCC is evident. Notably, an escalation in the SC from 75 to 175 g/L corresponded with an increase in the UCS from 0.63 MPa to 1.62 MPa, respectively.

5.2 Chemical Conversion Efficiency

The Chemical Conversion Efficiency (CCE) of the EICP was primarily assessed based on the quantity of urea hydrolysed into ammonium ions. Figure 3 illustrates the impact of soybean concentration on the CCE of the EICP-cemented CBR samples. As delineated earlier, the treatment process involved alternating treatment cycles of EICP solution (enzyme + urea + calcium chloride) and cementation solution (urea + calcium chloride). In total, seven treatment cycles were applied including four cycles with EICP solution (denoted in Fig. 3 as EICP1-4) and three cycles in

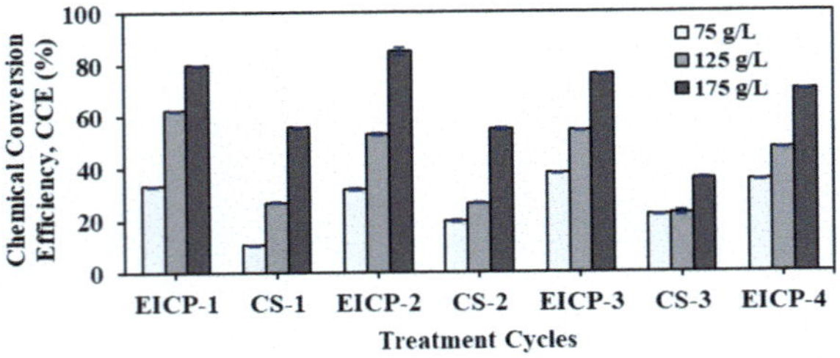

Fig. 3 Effect of soybean concentration of alternating treatment cycles on chemical conversion efficiency of EICP-cemented CBR samples

between with cementation solution alone (denoted in Fig. 3 as CS1-3). In general terms, an upward trend in the CCE was discernible with the increase in soybean concentration. Specifically, for the four EICP treatment cycles (EICP1-4) involving (enzyme + urea + calcium chloride), the CCE spanned between 32–38%, 48–62% and 70–85% correlating to 75, 125 and 175 g/L soybean concentrations, respectively. However, the CCE values for the CS treatment cycles (CS1-3) involving (urea + calcium chloride) were observed to be lower than those for the EICP treatment cycles, ranging between 11–22%, 23–27% and 36–56% correlating to 75, 125 and 175 g/L soybean concentrations, respectively. The relatively low CCE values for the CS treatment cycles can be attributed to the dependency of the reaction on the quantity of urease retained in the samples. The retention of urease might transpire on the particle surfaces of the CRB samples due to either adsorption, filtration or preservation of urease within the precipitated calcium carbonate, as suggested by Cui et al. [17].

5.3 *Comparison Between EICP-Treated and OPC-Treated CRB Samples*

Figure 4 illustrates the unconfined compressive strength (UCS) comparison between Enzyme-Induced Carbonate Precipitation (EICP) and Ordinary Portland Cement (OPC) treatments applied to crushed rock base (CRB) material. According to Austroads guides [34], a stabilised road base can be categorised into three cemented materials, including modified (UCS < 1.0 MPa), lightly bound (1.0 MPa $\leq$ UCS $\leq$ 2.0 MPa) and bound (UCS > 2.0 MPa). Based on the results obtained from the current study, the UCS of the EICP-treated CRB samples with 75 g/L treatment commensurate with the Austroads requirements for the modified road base material. Conversely, the results also indicated that the UCS of the EICP-treated CBR-treated samples with 125 and 175 g/L treatments are categorised as lightly bound cemented materials. In contrast, the OPC-treated CRB samples exhibit higher UCS values than those treated with EICP. As such, they qualify as bound cemented materials according to Austroads guides, with UCS values exceeding 2 MPa. This distinction underscores the variance in UCS between EICP-treated and OPC-treated samples for bound road

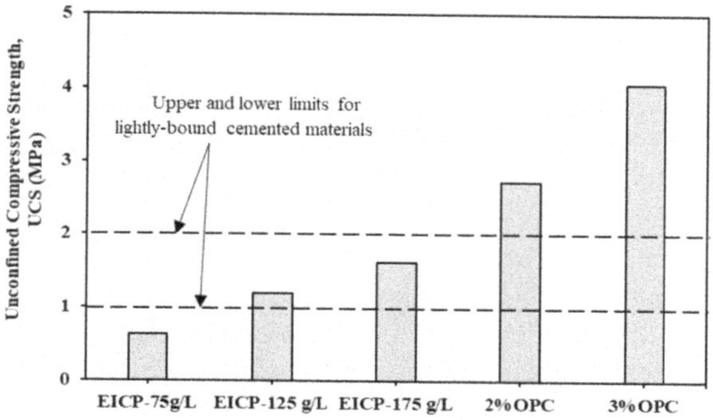

Fig. 4 Comparison between UCS of EICP-treated and OPC-treated CRB samples

base material, further reinforcing the need for careful material selection in the stabilisation processes. However, it is also obvious that the EICP-treated CBR samples were also successful in achieving the Austroads guides for lightly bound road base materials, reinforcing the importance of this technique for future road construction.

6 Conclusion

This paper presented an exploration of bio-cementation using Enzyme-Induced Carbonate Precipitation (EICP) for soil stabilisation, with special reference to transportation infrastructure, particularly in the stabilisation of crushed rock base (CRB) material. It was evident from the literature that this technology, which leverages earth-based bacteria or microorganisms for biological soil stabilisation, has rapidly gained prominence as a green alternative to traditional soil improvement using chemical additives such as lime or Portland cement. However, it was also recognised that further research is needed in this area, given the relative lack of studies focusing on coarse aggregates, the principal material used for base and subbase layers of pavements. A preliminary experimental study was conducted in this paper and demonstrated the effectiveness of EICP in enhancing the unconfined compressive strength of CRB for road applications, and research is currently ongoing by the authors to fully investigate the effectiveness of this promising stabilisation technology for road base materials in terms of mitigation of tensile failure, durability, long-term performance and economic viability. The following specific conclusions were drawn from the limited experimental results carried out in this paper:

- Significant increases in the unconfined compressive strength (UCS) and calcium carbonate content (CCC) were noted in the EICP-treated CRB samples when the soybean concentration (SC) varied from 75 to 175 g/L. A low SC of 75 g/L resulted in a final urease activity of 4 U/mL, which was too low to hydrolyse 50% of the urea within a 24 h reaction time. On the contrary, higher SC of 125 g/L and

175 g/L resulted in high urease activity (15 and 20 U/mL respectively), sufficient to hydrolyse approximately 54% and 78%, respectively, of urea within 24 h.
- The strength of the EICP-stabilised CRB samples showed a high dependency on the calcium carbonate content (CCC) precipitated within the pore spaces.
- The CRB samples stabilised with 2 and 3% Ordinary Portland Cement (OPC) exhibited strength greater than 2 MPa and are thus classified as bound cemented materials according to the Austroads guides. On the other hand, the EICP-stabilised CRB samples having SC of 125 and 175 g/L had UCS values greater than 1 MPa but less than 2 MPa and fell within the limits of lightly bound cemented materials according to the Austroads guides. This study thus underscores the potential of bio-cementation via EICP in achieving results comparable to OPC for bound CRB, however, for lightly bound CRB, the EICP supersedes OPC in avoiding the environmental repercussions and shrinkage crack issues associated with OPC hydration. In addition, research is currently ongoing to obtain EICP-cemented CRB comparable to OPC-cemented CRB for bound material.

References

1. Mitchell JK, Soga K (2005) Fundamentals of soil behaviour, vol 3. Wiley, New York
2. Bergado DT, Anderson LR, Miura N, Balasubramaniam AS (1996) Soft ground improvement in lowland and other environments. American Society of Civil Engineers, New York, United States of America
3. Schneider M (2019) The cement industry on the way to a low-carbon future. Cem Concr Res 124
4. Song JY, Kim Y, Jang J, Yun TS, Sim Y (2018) Microstructure of bio-mediated sand by enzyme induced carbonate precipitation: relevance to physio-mechanical properties. In: Proceedings of the 7th international conference on unsaturated soils, Hong Kong, China, pp 3–5
5. Cheng L, Shahin MA (2019) Microbially induced calcite precipitation (MICP) for soil stabilization. In: Ecological wisdom inspired restoration engineering
6. Mujah D, Cheng L, Shahin MA (2019) Microstructural and geomechanical study on bio-cemented sand for optimization of MICP process. J Mater Civ Eng 31(4)
7. Holten-Andersen N, Jaishankar A, Harrington MJ, Fullenkamp DE, DiMarco G, He L, McKinley GH, Messersmith PB, Lee KYC (2014) Metal-coordination: using one of nature's tricks to control soft material mechanics. J Mater Chem B 2(17)
8. Kavazanjian Jr E, Karatas I (2008) Microbiological improvement of the physical properties of soil. In: International conference on case histories in geotechnical engineering
9. Mujah D, Shahin MA, Cheng L (2017) State-of-the-art review of biocementation by microbially induced calcite precipitation (MICP) for soil stabilization. Geomicrobiol J 34(6)
10. Shahin MA, Jamieson K, Cheng L (2020) Microbial-induced carbonate precipitation for coastal erosion mitigation of sandy slopes. Géotech Lett 10(2)
11. Hamed Khodadadi T, Kavazanjian E, van Paassen L, DeJong J (2017) Bio-grout materials: a review. In: Grouting 2017
12. Cheng L, Shahin MA, Chu J (2019) Soil bio-cementation using a new one-phase low-pH injection method. Acta Geotech 14
13. Li M, Wen K, Li Y, Zhu L (2018) Impact of oxygen availability on microbially induced calcite precipitation (MICP) treatment. Geomicrobiol J 35(1)
14. Nafisi A, Safavizadeh S, Montoya BM (2019) Influence of microbe and enzyme-induced treatments on cemented sand shear response. J Geotech Geoenviron Eng 145(9)

15. Hoang TP (2018) Sand and silty-sand soil stabilization using bacterial enzyme-induced carbonate precipitation (BEICP). Ph.D. dissertation, Iowa State University
16. Amini Kiasari M, Pakbaz MS, Ghezelbash GR (2019) Comparison of effects of different nutrients on stimulating indigenous soil bacteria for biocementation. J Mater Civ Eng 31(6)
17. Cui M-J, Lai H-J, Hoang T, Chu J (2021) One-phase-low-pH enzyme-induced carbonate precipitation (EICP) method for soil improvement. Acta Geotech 16
18. Arab MG, Alsodi R, Almajed A, Yasuhara H, Zeiada W, Shahin MA (2021) State-of-the-art review of enzyme-induced calcite precipitation (EICP) for ground improvement: applications and prospects. Geosciences 11(12)
19. Zhao Q, Li L, Li C, Li M, Amini F, Zhang H (2014) Factors affecting the improvement of engineering properties of MICP-treated soil catalyzed by bacteria and urease. J Mater Civ Eng 26(12)
20. Krajewska B (2018) Urease-aided calcium carbonate mineralization for engineering applications: a review. J Adv Res 13
21. Almajed A, Tirkolaei HK, Kavazanjian Jr E (2018) Baseline investigation on enzyme-induced calcium carbonate precipitation. J Geotech Geoenviron Eng 144(11)
22. Nemati M, Voordouw G (2003) Modification of porous media permeability, using calcium carbonate produced enzymatically in situ. Enzyme Microb Technol 33(5)
23. Rebata-Landa V (2007) Microbial activity in sediments: effects on soil behaviour. Ph.D. dissertation, Georgia Institute of Technology, Atlanta, United States of America
24. Chouhan S, Vishnu Priya V, Gayathri R (2018) Extraction and partial purification of urease enzyme from jack fruit. Int J Res Pharm Sci 9(2)
25. Xiao P et al (2019) Effect of relative density and biocementation on the cyclic response of calcareous sand. Can Geotech J 56(12)
26. Nafisi A, Montoya BM, Matthew Evans T (2020) Shear-strength envelopes of bio-cemented sands with varying particle size and cementation levels. J Geotech Geoenviron Eng 146(3)
27. Xiao Y et al (2019) Effect of particle shape on strength and stiffness of bio-cemented glass beads. J Geotech Geoenviron Eng 145(11)
28. Chandra A, Ravi K (2021) Application of enzyme-induced carbonate precipitation (EICP) to improve the shear strength of different types of soils. In: Problematic soils and geoenvironmental concerns: proceedings of IGC 2018. Springer Singapore, Singapore
29. Oliveira PJV, Freitas LD, Carmona JPSF (2017) Effect of soil type on the enzymatic calcium carbonate precipitation process used for soil improvement. J Mater Civ Eng 29(4)
30. Hamdan D (2015) Applications of enzyme-induced carbonate precipitation (EICP) for soil improvement. Ph.D. dissertation, Arizona State University, Tempe, United States of America
31. Kavazanjian Jr E, Almajed A, Hamdan N (2017) Bio-inspired soil improvement using EICP soil columns and soil nails. In: Grouting 2017, pp 13–22
32. Neupane D, Yasuhara H, Kinoshita N, Unno T (2013) Applicability of enzymatic calcium carbonate precipitation as a soil-strengthening technique. J Geotech Geoenviron Eng 139(12)
33. Knorr B (2014) Enzyme-induced carbonate precipitation for the mitigation of fugitive dust. Master dissertation, Arizona State University, Tempe, United States of America
34. Austroads (2019) Guide to pavement technology part 4D: stabilised materials (Austroads project no. APT6105). Austroads (publication no. AGPT04D-19), Sydney
35. Austroads (2017) Guide to pavement technology: part 2
36. Martin KK, Khodadadi TH, Chester M, Kavazanjian Jr E (2020) Hotspot life cycle assessment for environmental impacts of EICP for ground improvement. In: Geo-congress 2020. American Society of Civil Engineers, Reston, VA, pp 321–329
37. Pasillas JN, Khodadadi H, Martin K, Bandini P, Newtson CM, Kavazanjian Jr E (2018) Viscosity-enhanced EICP treatment of soil. In: IFCEE 2018, pp 145–154
38. Australian Standard (2009) Methods of testing soils for engineering purposes method 3.6.1: soil classification tests—determination of the particle size distribution of a soil—standard method of analysis by sieving, AS1289.3.6.1-2009
39. Main Roads Western Australia (2019) WA 115.2—particle size distribution abbreviated method for coarse and medium-grained soils test method, Perth, WA, Australia

40. Whiffin VS (2004) Microbial $CaCO_3$ precipitation for the production of biocement. Ph.D. dissertation, Murdoch University, Perth, WA, Australia
41. Australian Standard (2008) Methods for preparation and testing of stabilized materials method 4: unconfined compressive strength of compacted materials, AS5101.4-2008

Installation Effects on the Capacity of Open-Ended and Closed-Ended Pipe Piles

Monica Prezzi, Venkata Abhishek Sakleshpur, and Daniel G. Fridman

Abstract The base and shaft capacities of piles depend on the method of pile installation, pile type, soil type, and stress state. Fully instrumented, open-ended, and closed-ended pipe piles were load tested side-by-side statically following a slow-maintained testing protocol at two sites in Indiana, USA. The piles were instrumented following a hybrid, dense instrumentation scheme with both vibrating-wire and electrical-resistance strain gauges along the entire length of the shafts. The open-ended piles were specially designed and assembled following a double-wall system to facilitate separation of the external and internal unit shaft resistances in every layer of the soil profile. The plug length ratio (PLR) and incremental filling ratio (IFR) were measured for the open-ended test piles. To gain insights on the impact of the formation and evolution of the soil plug during installation on the base capacity of pipe piles, one open-ended model pile and one closed-ended model pile were installed under similar conditions in a calibration chamber with digital image capabilities. This paper summarizes and discusses the results of the pile load tests. The load test results are compared with predictions obtained using modern CPT-based pile design methods available in the literature. An assessment of the design methods is made based on the field test results.

Keywords Pile load test · Open-ended pile · Closed-ended pile · DIC

1 Introduction

Piles are generally categorized as non-displacement or displacement piles, depending on the installation method. A closed-ended pipe pile (CEP) is a displacement pile, whereas the response of an open-ended pipe pile (OEP) typically lies between that of a displacement and a non-displacement pile [1]. A soil plug develops inside an OEP during installation [2], and the plug length L_p may be less than or equal to the

M. Prezzi (✉) · V. A. Sakleshpur · D. G. Fridman
Lyles School of Civil Engineering, Purdue University, West Lafayette, IN 47907, USA
e-mail: mprezzi@ecn.purdue.edu

B. Indraratna and C. Rujikiatkamjorn (eds.), *Recent Advances and Innovative Developments in Transportation Geotechnics*,
https://doi.org/10.1007/978-981-97-8245-1_2

pile penetration depth L. If $L_p = L$, the pile is driven in an unplugged/fully coring mode throughout, whereas if $L_p < L$, the pile is driven in a partially or fully plugged mode at least during part of the installation [3]. In general, when comparing the response of pipe piles of the same geometry installed in the same soil profile, an OEP driven in a fully coring mode would have a load response that is stiffer than that of a non-displacement pile, whereas an OEP driven in a fully plugged mode would have a load response that approaches that of a CEP [4].

Side-by-side field static load tests (SLTs) on fully instrumented CEPs and OEPs installed in the same soil profile allow us to study the effects of pile installation on the pile shaft and base capacities. Calibration chamber tests with image-capturing capabilities provide the means to systematically study the response of pipe piles during installation and loading under controlled conditions. This paper summarizes the results of side-by-side SLTs performed on full-scale, instrumented CEPs and OEPs at two sites in Indiana, USA. The test results are compared with predictions obtained using modern CPT-based pile design methods. We also present preliminary results of a model OEP and a model CEP tested in the digital image correlation (DIC) calibration chamber at Purdue University. The effects of pile installation are reflected in the measured displacement fields in the soil domain surrounding the model piles.

2 Pile Load Tests in Indiana

2.1 Lagrange County

A CEP and an OEP were tested side-by-side at a bridge construction site over the Pigeon River in Lagrange County, Indiana [5, 6]. The soil profile consists of gravelly sand (Table 1); the top 3 m is in a loose state ($D_R \approx 30\%$), while the rest of the deposit is in a dense-to-very dense state ($D_R \approx 80\%$), with SPT blow counts ranging from 20 to 60, and cone resistance q_c values, from 15 to 25 MPa. The water table is at 3 m depth below the ground surface.

The wall thickness t and outer diameter B of the CEP are equal to 12.7 and 356 mm. The double-wall OEP was assembled by combining an inner pipe with B = 305 mm and an outer pipe with B = 356 mm; both pipes had the same value of

Table 1 Soil layers at the pile load test site in Lagrange County, Indiana [5, 6]

z (m)	Layer description	γ_m (kN/m^3)	e_{min} (–)	e_{max} (–)	D_R (%)	ϕ_c (°)
0–3	Gravelly sand	16.38	0.41	0.68	30	33.3
3–13	Gravelly sand	21.20	0.41	0.68	80	33.3

Note z = depth below ground surface, e_{min} = minimum void ratio, γ_m = unit weight, e_{max} = maximum void ratio, ϕ_c = critical-state friction angle, and D_R = relative density

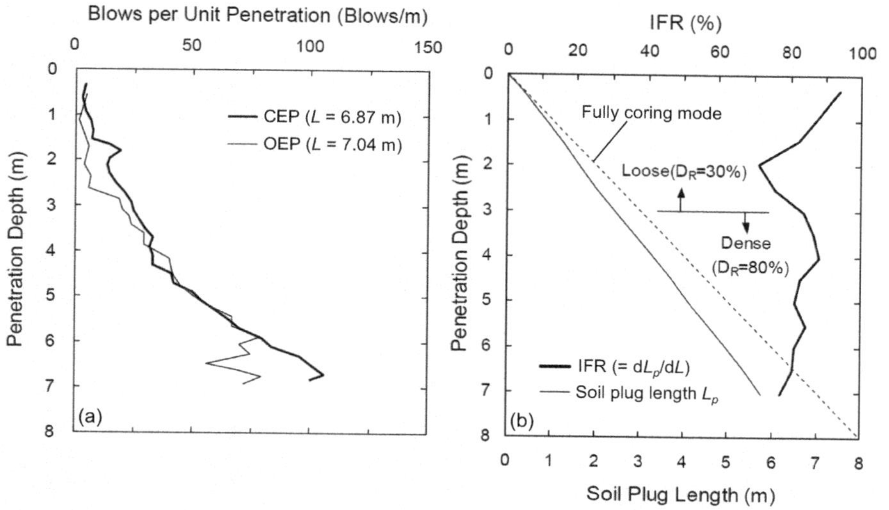

Fig. 1 Pile driving results for the Lagrange County site (modified after [5]): **a** driving resistance (blow counts) of CEP and OEP, and **b** IFR and soil plug length for the OEP

t (= 6.4 mm). The spacing between the OEP and the CEP was about 4.9 m. Nine pairs of electrical-resistance (ER) strain gauges were installed on the CEP, while the inner and outer pipes of the OEP were instrumented with ten pairs and nine pairs of ER strain gauges, respectively.

The CEP and the OEP were driven to depths of 6.87 and 7.04 m (Fig. 1a). Figure 1b shows that the OEP was partially plugged and never reached a fully plugged state (with IFR = 0; IFR = dL_p/dL). The values of PLR (= L_p/L) and IFR at the end of OEP driving were 0.82 and 0.78. Measurements of soil plug length recorded before and after the SLT showed that the soil plug length did not change as a result of the SLT, indicating that either the OEP behaved as a fully plugged pile under static loading, or any compression of the soil plug was likely compensated by soil entering the pile from the bottom during loading.

2.2 *Tippecanoe County*

A CEP and an OEP were tested side-by-side at a bridge construction site over the Wabash River in Tippecanoe County, Indiana [7, 8]. The soil profile at the site generally consists of layers of poorly graded sand with different gravel contents (Table 2). The gravel content was small (< 20%) for depths less than 15 m, except for a narrow layer at around 9 m depth, where the gravel content exceeded 50%. The gravel content ranged from 30 to 60% for depths exceeding 15 m. The water table was found at a depth of 3.05 m below the ground surface. Additional site investigation results are reported in [7, 8].

The wall thickness t and outer diameter B of the CEP are equal to 13 and 610 mm. The OEP was composed of a bottom segment and a top segment (each 18.3 m long). The bottom double-wall segment consists of an inner pipe with $B = 584$ mm and an outer pipe with $B = 660$ mm; both pipes had the same value of t (= 13 mm). The outer diameter of the bottom and top segments of the OEP were the same, but the value of t for the top segment was 19 mm. The spacing between the OEP and the CEP was about 7.3 m.

The shaft of the CEP was instrumented with 20 pairs of ER and 12 pairs of vibrating-wire (VW) strain gauges [9, 10]. For the double-wall OEP, 19 pairs of ER gauges were installed along the inner pipe of the bottom segment, whereas 11 and 14 pairs of VW and ER gauges were installed along the outer pipe of the bottom segment [11]. In addition, the top segment of the OEP was instrumented with three and five pairs of VW and ER gauges, respectively [12]. Further information about the CEP and OEP instrumentation is reported in the literature [9–12].

The CEP was driven to a depth of 17.37 m below the ground surface (Fig. 2a). For installation of the OEP, a 0.9 m-diameter steel casing was first installed up to a depth of 8.53 m, the soil inside the casing was excavated, and then the OEP was driven to a depth of 30.48 m below the ground surface. Figure 2b shows that the IFR decreased from 0.92 to 0.70 during installation of the OEP, and that the PLR was approximately 0.78 at the end of driving of the OEP. The OEP and the CEP were load tested 8 days and 13 days, respectively, after pile driving.

Table 2 Soil layers at the pile load test site in Tippecanoe County, Indiana [7, 8]

z (m)	Layer description	γ_m (kN/m^3)	D_{50} (mm)	GC (%)	C_U (–)	ϕ_c (°)	δ_c (°)
0.0–5.5	Clayey silt with sand	19.5	–	0	3.0	30	24.6
5.5–8.2	Sand with gravel	20.0	0.4	4	2.6	32	26.2
8.2–10.4	Sandy gravel	21.5	4.5	49	34.6	35	23.1
10.4–16.8	Sand with gravel	20.0	0.9	10	4.8	32	24.3
16.8–22.6	Gravelly sand	21.5	4.1	43	16.6	34	23.1
22.6–32.6	Gravelly sand	21.5	1.1	28	8.3	33	25.1

Note z = depth below ground surface, D_{50} = mean particle size, γ_m = unit weight, GC = gravel content (= percent of soil particles by mass retained on the 4.75 mm sieve), ϕ_c = critical-state friction angle, C_U = coefficient of uniformity, and δ_c = critical-state interface friction angle

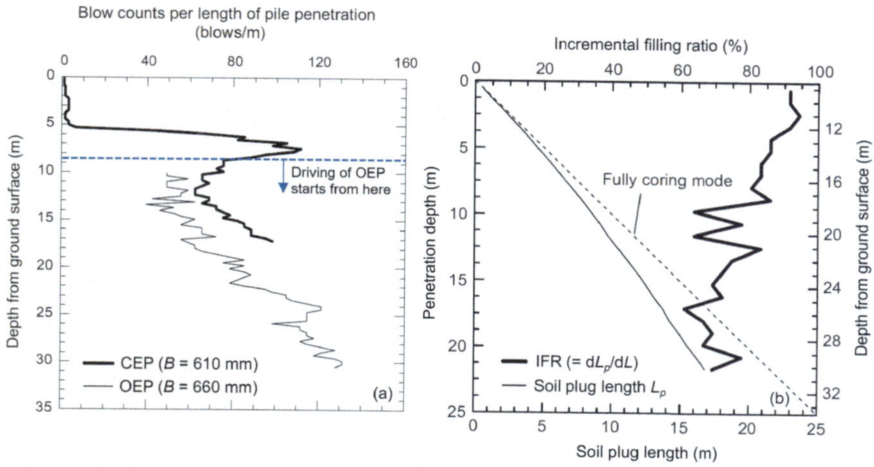

Fig. 2 Pile driving results for Tippecanoe County site: **a** comparison between driving resistances of CEP and OEP (after [7]), and **b** soil plug length and IFR for OEP (after [11])

3 Comparison Between Predicted and Measured Resistances

Several CPT-based methods have been proposed to predict the ultimate load capacity of CEPs and OEPs in sand [13–16]. Some of these methods include the Purdue Pile Design Method (PPDM) [7, 17], the Unified Pile Design Method (UPDM) [18], and the Imperial College Pile Design Method (ICPDM) [19].

Figure 3 compares the predicted and measured axial load transfer curves at the ultimate load level (when $w = 0.1B$) for the instrumented CEPs and OEPs tested in Indiana; w = pile head settlement. The axial load transfer curve for the OEP in Lagrange County was not included in Fig. 3 because some of the gauges in the lower part of the outer pipe malfunctioned after pile driving [5]. The PPDM, ICPDM, and UPDM provided good estimates of the base and shaft capacities of the CEP in Lagrange County (Fig. 3a). For the CEP in Tippecanoe County, all three methods estimated the limit shaft capacity Q_{sL} quite well—the PPDM, ICPDM, and UPDM provided estimates of Q_{sL} that were within ± 10% of the measured value, with the ICPDM just 5% below the measured shaft capacity. However, the methods overpredicted the ultimate base capacity $Q_{b,ult}$ of the CEP due to the high gravel content (= 50%) found at the pile base level, resulting in a large q_c value (= 28.7 MPa, averaged from 1B above to 2B below the pile base) for use in the pile base capacity calculations. The ICPDM provided the closest estimate of $Q_{b,ult}$ (= 1.5 times the measured base capacity) for the CEP in Tippecanoe County (Fig. 3b). For the OEP in Tippecanoe County, we see the opposite trend, where all three methods predicted the pile base capacity reasonably well—the PPDM, ICPDM, and UPDM provided estimates of $Q_{b,ult}$ that were within − 12% of the measured value, with the PPDM

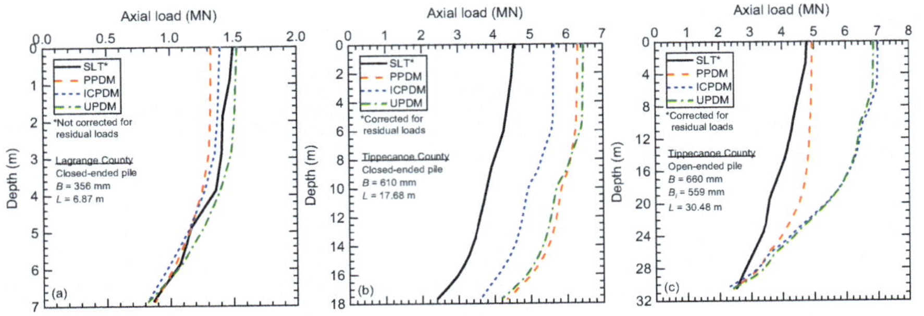

Fig. 3 Predicted and measured axial load transfer curves at the ultimate load level for **a** CEP in Lagrange County, **b** CEP in Tippecanoe County, and **c** OEP in Tippecanoe County

just 3% below the measured base capacity, even though the gravel content at the level of the OEP base was 28%. This is because: (1) the annulus thickness of the OEP (= 51 mm) was comparable to the cone diameter d_c (= 44.5 mm), resulting in a cone resistance that was representative of the resistance mobilized by the pile annulus, and (2) the measured annulus resistance Q_{ann} was approximately 88% of the base resistance $Q_{b,ult}$ of the OEP at the ultimate load level. While the PPDM predicted the limit shaft capacity of the OEP quite well, both the ICPDM and UPDM overpredicted Q_{sL} by a factor of two (Fig. 3c). A layer-by-layer comparison of the predicted and measured limit unit shaft resistance q_{sL} profiles in Tippecanoe County indicates that the overestimation of q_{sL} for the CPT-based methods occurred at depths where the gravel content was higher than 30% [7, 11]. Although the gravel content was not reported for the Lagrange County test site, the fairly good match between the measured and predicted axial load transfer curves suggests that the gravel content at this site was probably low (< 30%).

In practice, when detailed site characterization is unavailable, engineers often rely on parameters c_s (= q_{sL}/q_c) and β (= q_{sL}/σ'_{v0}) to calculate the outer unit shaft resistance q_{sL} of pipe piles in sand; σ'_{v0} = in situ vertical effective stress at the elevation where q_{sL} is computed. Figure 4 shows the values of c_s and β determined from the SLT results at the ultimate load level for the instrumented CEPs and OEPs in Indiana. For both test sites, the values of c_s and β are generally higher for the CEP than the OEP; this is likely because of the displacement of a greater volume of the surrounding soil by the driving of CEPs than OEPs, resulting in a higher degree of soil densification and lateral stress increase around the shaft of the CEPs. The c_s and β values determined from the SLT data generally agree with the range of values documented in the literature (c_s = 0.1–0.9% and β = 0.3–0.9 for driven piles in sand [1, 17, 20]).

Figure 4 also shows the values of the lateral earth pressure coefficient K mobilized within the depth range of the active plug length for the instrumented OEPs in Indiana. The magnitude and distribution of K_{inner} and K_{outer}, which are the values of K (= $q_{sL}/$

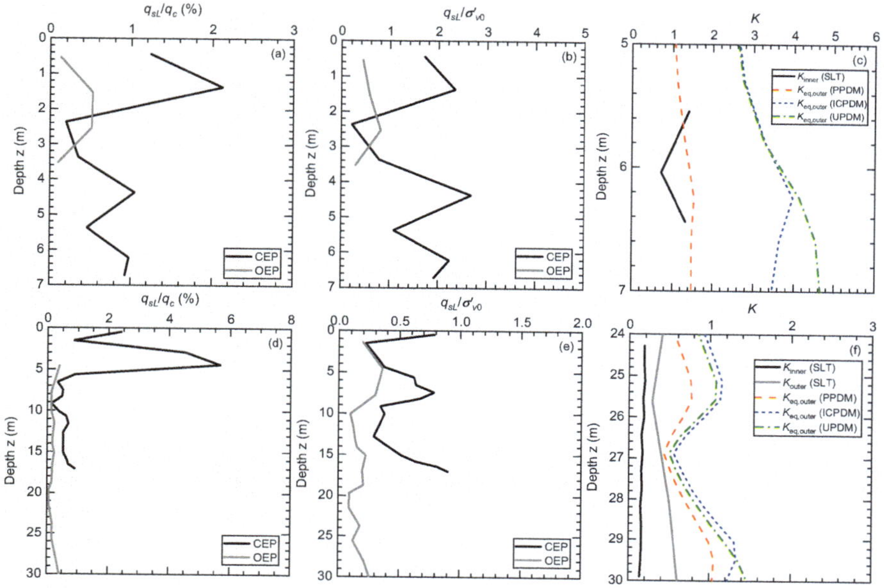

Fig. 4 Profiles of c_s ($= q_{sL}/q_c$), β ($= q_{sL}/\sigma'_{v0}$), and K ($= q_{sL}/\sigma'_{v0} \tan \delta_c$) for pipe piles installed in gravelly sand deposits in Lagrange County (**a**–**c**) and Tippecanoe County (**d**–**f**)

$\sigma'_{v0} \tan \delta_c$) mobilized along the inner and outer pipes of the double-wall OEPs, were determined at the ultimate load level from the strain gauges installed on the surfaces of the inner and outer pipes. The value of K_{inner} was determined by assuming that the vertical effective stress σ'_v inside the plug is equal to the value of σ'_{v0} outside the pile at the elevation of interest. For the CPT-based design methods, the magnitude and distribution of $K_{eq,outer}$, which is the equivalent K mobilized along the outer pipe of the OEP, was estimated from the method equations; for example, $K_{eq,outer} = K(1 - 0.66\text{PLR})$ for the PPDM, and $K_{eq,outer} = (\sigma'_{rc} + \Delta\sigma'_{rd})/\sigma'_{v0}$ for both the ICPDM and UPDM (using the expressions for K, σ'_{rc}, and $\Delta\sigma'_{rd}$ as defined in the methods). Figure 4c shows that, for the OEP in Lagrange County, the value of K_{inner} is approximately equal to 1.0, on average, while the values of $K_{eq,outer}$ range from 1.0 to 4.6, with smaller values corresponding to the PPDM. Similarly, Fig. 4f shows that, for the OEP in Tippecanoe County, the values of K_{inner} and K_{outer} range from about 0.14 to 0.16 and 0.3 to 0.6, while the values of $K_{eq,outer}$ range from 0.4 to 1.4, with smaller values corresponding to the PPDM. The values of K_{inner} from the SLT are approximately constant within the active plug length for the OEP in Tippecanoe County. The active plug length $L_{p,A}$ at the ultimate load level is approximately equal to 7 and 11 inner diameters for the OEPs in Lagrange County and Tippecanoe County, respectively. The ratio $L_{p,A}/L_p$ of the active plug length to the total plug length (measured at the end of driving) was comparable for the OEPs at both test sites (= 35 and 38% for Lagrange and Tippecanoe Counties).

For the OEP in Tippecanoe County, the value of β inside the plug was calculated at the ultimate load level by assuming that σ'_{v} inside the plug is equal to σ'_{v0} outside the pile at the elevation of interest. This is a practical approach because, in design, the vertical effective stresses inside the plug are unknown. Following this approach, the values of β obtained from the SLT along the active plug length range from about 0.06 to 0.08. Using the formulation derived in [21], which assumes that β is constant along the active plug length, we calculated a value of β equal to 0.05 by assuming that the vertical effective stress at the base of the plug is equal to the unit plug resistance q_{plug} (= 1.3 MPa) obtained from the SLT at the ultimate load level. For OEPs driven under partially plugged conditions, the practical approach of using σ'_{v0} is convenient in the calculation of unit plug resistance, and β values within the above range may be used as reference for OEPs installed in soil profiles similar to that at the Tippecanoe County test site. Note that, for smaller values of IFR, with driving approaching fully plugged conditions, the β values may vary along the active plug length. In the formulation of [21], the vertical effective stresses in the soil plug increase exponentially near the pile base, but this change in the vertical effective stress may be due in part to these varying β values. Calibration chamber test data indicate that the transfer of load in the soil plug tends to increase significantly near the pile base [2, 22]. Another factor in the experimental determination of β values inside the plug is the separation of the annulus resistance from the plug resistance based on discrete instrumentation within the inner pipe wall.

Table 3 summarizes the measured and predicted values of the normalized ultimate unit base resistance c_{b} (= $q_{b,\text{ult}}/q_{cb}$) for the instrumented CEPs and OEPs tested in gravelly sand profiles in Indiana; q_{cb} = representative cone resistance at the pile base level. For the SLT $q_{b,\text{ult}}/q_{cb}$ ratios, the value of q_{cb} was calculated by taking the average of the q_{c} values over a distance of $1B$ above and $2B$ below the pile base. For Lagrange County, the CPT-based methods predict values of c_{b} that are in good agreement with those measured from the SLTs for both the CEP and OEP. For Tippecanoe County, the predicted and measured c_{b} values agree closely with each other for the OEP but not for the CEP (for the reasons mentioned previously). Given that the limit unit base resistance q_{bL} (at plunging) of a CEP, particularly one that has a conical tip (as is the case for the CEP tested in Tippecanoe County), is generally comparable to the cone resistance measured near the pile base [17], the values of c_{b} were recalculated assuming q_{cb} to be equal to q_{bL} (= 10.4 MPa, measured in the SLT on the CEP). The values of c_{b} estimated from the CPT-based design methods, assuming $q_{cb} = q_{bL}$, tend to be closer to the measured value (= 0.79), except for the UPDM, where c_{b} is always equal to 0.50 for CEPs in sand. The unit plug resistance q_{plug} and the unit annulus resistance q_{ann} of the OEP in Tippecanoe County at the ultimate load level were 6 and 93% of q_{cb} (= 22.8 MPa, averaged $1B$ above and $2B$ below the pile base).

Table 3 Predicted and measured values of c_b (= $q_{b,ult}/q_{cb}$) for CEPs and OEPs tested in Indiana

Test/design method	$q_{b,ult}/q_{cb}$ [Lagrange]		$q_{b,ult}/q_{cb}$ [Tippecanoe]		
	CEP	OEP	CEP[b]	CEP[c]	OEP
SLT ($w = 0.1B$)	0.55[a]	0.35[a]	0.29	0.79	0.32
PPDM	0.56	0.28	0.52	0.77	0.31
ICPDM	0.50	0.33	0.43	1.00	0.28
UPDM	0.50	0.30	0.50	0.50	0.31

[a] Not corrected for residual loads (due to drifts in the zero readings of the strain gauges)
[b] Based on q_{cb} calculated using the CPT data
[c] Based on q_{cb} assumed as equal to q_{bL} (= 10.4 MPa) measured from the SLT

4 DIC-Based Calibration Chamber Tests

The digital image correlation (DIC) technique can be used to obtain the displacement and strain fields in the soil domain around model footings [23–26] and model cones/piles [27–37] tested in calibration chambers with image capabilities. The DIC chamber at Purdue University is a half-cylindrical calibration chamber with observation windows positioned on its plane of symmetry. All boundaries of the chamber are fixed-wall boundaries, except the top surface of the soil sample, where a vertical stress (surcharge) can be applied. Additional details about the chamber and the DIC technique can be found in [27]. Tests were performed on a model OEP and CEP installed in uniform dense samples ($D_R \approx 90\%$) of Ottawa 20–30 (OTC) sand (a clean silica sand with $D_{50} = 0.72$ mm and $\phi_c = 29.3°$). The steel model OEP has an outer diameter of 50.8 mm and a wall thickness of 9.5 mm, while the conical-tip brass model CEP has a diameter of 38.1 mm.

The sand samples were air-pluviated inside the chamber, and the top surface of the sand was carefully leveled to facilitate application of the vertical stress using an inflatable air bladder. For both the model CEP and OEP tests, the vertical stress at the center of the second observation window (end of the test) was approximately 40 kPa. The model CEP and OEP were preinstalled 50 and 140 mm into the sand sample (before application of the surcharge) to ensure pile alignment and prevent sand intrusion between the observation window and the model pile. The piles were monotonically jacked into the sand at a rate of 1 mm/s. During the experiments, digital images of the sand domain were captured by two 12-megapixel monochrome cameras. The IFR values measured during penetration of the model OEP were in the range of 0.7–0.9, with a slight decrease in IFR with increasing pile penetration depth L.

Figure 5 shows the radial displacement du_r and vertical displacement du_z fields obtained for the model CEP and OEP in dense OTC sand. The DIC analyses correspond to 234 mm of pile penetration, from $L = 140$ mm (reference image) to $L = 374$ mm (end of test); the figure axes are normalized with respect to the pile radius r_p. The displacement fields for both the model CEP and OEP have a similar pattern, in the sense that vertical displacements dominate within the zone immediately below the pile base, while radial displacements dominate within the zone radially adjacent to the pile shafts. The radial displacement contours extend further into the sand domain for the model OEP than the CEP, with the 1.5 mm du_r contour, for example, reaching a vertical distance of $4r_p$ below the OEP base and a radial distance of $7r_p$ from the OEP centerline (Fig. 5a). A similar trend can also be observed for the vertical displacement fields; for example, the 2 mm du_z contour for the model OEP extends beyond a vertical distance of $5r_p$ below the pile base and up to a radial distance of $5r_p$ from the pile centerline (Fig. 5c). For the model CEP, the 2 mm du_z contour extends up to a vertical distance of $3.5r_p$ below the pile base and a radial distance of approximately $3r_p$ from the pile centerline (Fig. 5d). Despite the relatively high IFR values measured for the OEP, the model OEP produced greater overall soil displacements than the CEP, likely because of its larger diameter.

5 Conclusions

The paper summarizes two case histories of static load tests on fully instrumented, closed-ended pipe piles (CEPs) and double-wall open-ended pipe piles (OEPs) tested side-by-side in gravelly sand deposits in Indiana. Three CPT-based pile design methods were used to predict the axial load transfer curves of the test piles. Profiles of commonly used design parameters c_s, β, and K were developed from the SLT results; these profiles may be used as reference for calculation of unit shaft resistances of CEPs and OEPs installed in similar soil profiles. From the SLT results, the values of K mobilized along the inner pipe of the OEP at the ultimate limit state were compared with those mobilized along the outer pipe as well as those predicted by the design methods. The results of DIC-based calibration chamber tests on model CEPs and OEPs highlight the effects of pile installation on the magnitude and extent of the vertical and radial displacement fields in the soil domain around the model piles; these differences directly affect the load-transfer behavior and load-carrying capacity of these piles.

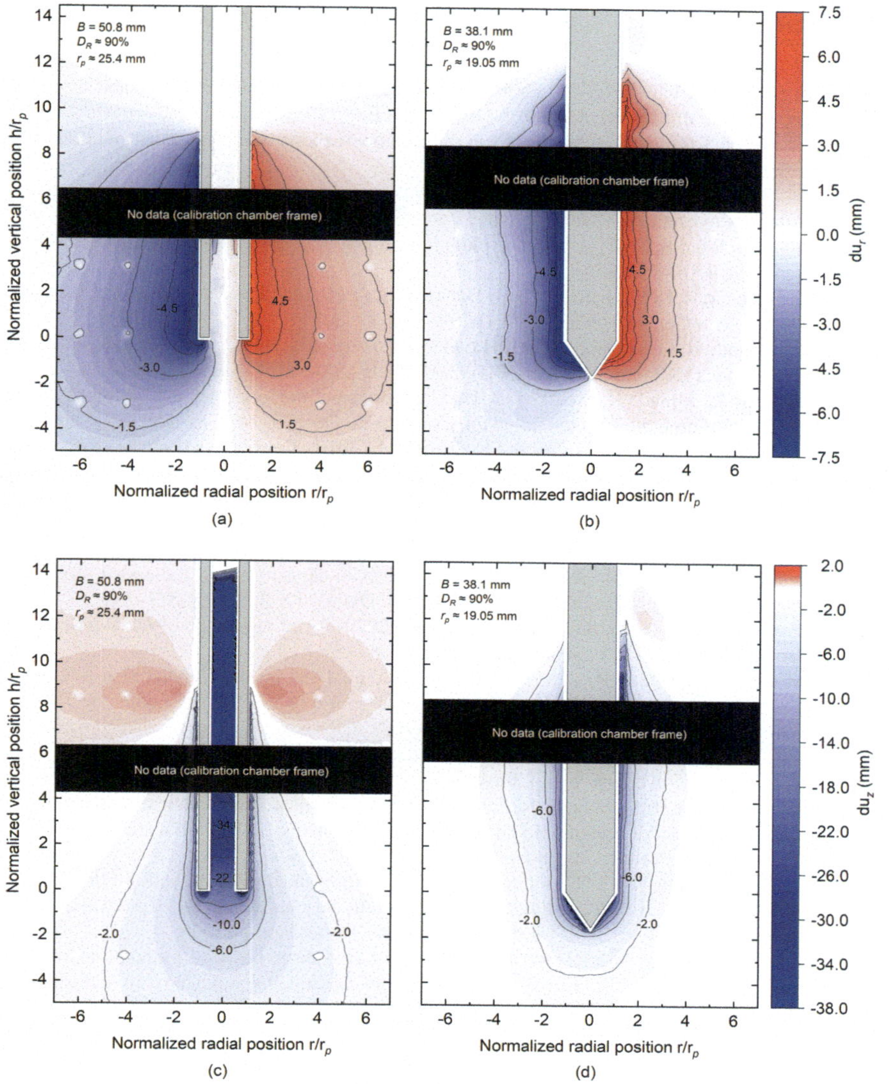

Fig. 5 DIC results for the model OEP and CEP in dense OTC sand showing the radial displacement fields (**a**, **b**) and the vertical displacement fields (**c**, **d**)

Acknowledgements The work related to the field pile load tests was funded by the Joint Transportation Research Program administered by the Indiana Department of Transportation and Purdue University through Contracts SPR-2361 and SPR-4040. The DIC calibration chamber research work was funded by the National Science Foundation under Grant No. 2028672. Any opinions, findings, and conclusions or recommendations expressed in this material are those of the author(s) and do not necessarily reflect the views of the National Science Foundation.

References

1. Lee J, Salgado R, Paik K (2003) Estimation of load capacity of pipe piles in sand based on cone penetration test results. J Geotech Geoenviron Eng 129(5):391–403
2. Paik K-H, Lee S-R (1993) Behavior of soil plugs in open-ended model piles driven into sands. Mar Georesour Geotechnol 11(4):353–373
3. Paik K, Salgado R (2003) Determination of bearing capacity of open-ended piles in sand. J Geotech Geoenviron Eng 129(1):46–57
4. De Nicola A, Randolph MF (1997) The plugging behaviour of driven and jacked piles in sand. Géotechnique 47(4):841–856
5. Paik K, Salgado R, Lee J, Kim B (2003) Behavior of open- and closed-ended piles driven into sands. J Geotech Geoenviron Eng 129(4):296–306
6. Kim K, Salgado R, Lee J, Paik K (2002) Load tests on pipe piles for development of CPT-based design method. JTRP publication no. FHWA/IN/JTRP-2002/04. Purdue University
7. Han F, Ganju E, Salgado R, Prezzi M (2019) Comparison of the load response of closed-ended and open-ended pipe piles driven in gravelly sand. Acta Geotech 14:1785–1803
8. Han F, Ganju E, Salgado R, Prezzi M, Zaheer M (2019) Experimental study of the load response of large diameter closed-ended and open-ended pipe piles installed in alluvial soil. JTRP publication no. FHWA/IN/JTRP-2019/03. Purdue University, West Lafayette, IN, USA
9. Ganju E, Han F, Prezzi M, Salgado R (2020) Static capacity of closed-ended pipe pile driven in gravelly sand. J Geotech Geoenviron Eng 146(4):04020008
10. Ganju E, Han F, Prezzi M, Salgado R (2021) The axial capacity of closed-ended pipe piles driven in gravelly sands. In: Proceedings of IFCEE. Dallas, TX, USA, pp 377–387
11. Han F, Ganju E, Prezzi M, Salgado R, Zaheer M (2020) Axial resistance of open-ended pipe pile driven in gravelly sand. Géotechnique 70(2):138–152
12. Han F, Ganju E, Salgado R, Prezzi M (2020) Static load test on open-ended pipe pile using double-wall instrumentation. In: Proceedings of geo-congress 2020. Minneapolis, pp 73–81
13. Sakleshpur VA, Prezzi M, Salgado R, Zaheer M (2021) CPT-based geotechnical design manual, volume 2: CPT-based design of foundations (methods). JTRP publication no. FHWA/IN/JTRP-2021/23. Purdue University, West Lafayette, IN, USA
14. Sakleshpur VA, Prezzi M, Salgado R, Zaheer M (2021) CPT-based geotechnical design manual, volume 3: CPT-based design of foundations (example problems). JTRP publication no. FHWA/IN/JTRP-2021/24. Purdue University, West Lafayette, IN, USA
15. Prezzi M, Sakleshpur VA (2022) Static and dynamic testing and design methods for driven piles in multilayered soil. In: Proceedings of 11th international conference on stress wave theory and design and testing methods for deep foundations. Rotterdam, pp 1–14
16. Sakleshpur VA, Prezzi M, Salgado R (2023) CPT-based design of pile foundations in sand and clay: perspectives. In: Proceedings of 8th international symposium on deformation characteristics of geomaterials. Porto, Portugal, pp 1–8
17. Salgado R (2022) The engineering of foundations, slopes and retaining structures, 2nd edn. CRC Press, Boca Raton, FL, USA
18. Lehane B, Liu Z, Bittar E, Nadim F, Lacasse S, Jardine R, Carotenuto P, Rattley M, Jeanjean P, Gavin K, Gilbert R, Bergan-Haavik J, Morgan N (2020) A new "unified" CPT-based axial pile capacity design method for driven piles in sand. In: Proceedings of 4th international symposium on frontiers in offshore geotechnics. Austin, TX, USA, pp 462–477
19. Jardine R, Chow F, Overy R, Standing J (2005) ICP design methods for driven piles in sands and clays. Thomas Telford Ltd., London, UK
20. Fellenius BH (2019) Basics of foundation design. Pile Buck International, Vero Beach, FL, USA
21. Randolph MF, Leong EC, Houlsby GT (1991) One-dimensional analysis of soil plugs in pipe piles. Géotechnique 41(4):587–598
22. Lehane BM, Gavin KG (2001) Base resistance of jacked pipe piles in sand. J Geotech Geoenviron Eng 127(6):473–480

23. Janabi FH, Sakleshpur VA, Prezzi M, Salgado R (2022) Strain influence diagrams for settlement estimation of square footings on layered sand. J Geotech Geoenviron Eng 148(5):04022025
24. Janabi FH, Raja RA, Sakleshpur VA, Prezzi M, Salgado R (2023) Experimental study of shape and depth factors and deformations of footings in sand. J Geotech Geoenviron Eng 149(2):04022128
25. Raja RA, Sakleshpur VA, Prezzi M, Salgado R (2023) Effect of relative density and particle morphology on the bearing capacity and collapse mechanism of strip footing in sand. J Geotech Geoenviron Eng 149(8):04023052
26. Raja RA, Sakleshpur VA, Prezzi M, Salgado R (2023) Load response and soil displacement field for a vertically loaded strip footing on sand underlain by a stiff base. J Geotech Geoenviron Eng 149(11):04023106
27. Arshad M, Tehrani F, Prezzi M, Salgado R (2014) Experimental study of cone penetration in silica sand using digital image correlation. Géotechnique 64(7):551–569
28. Tehrani FS, Han F, Salgado R, Prezzi M, Tovar RD, Castro AG (2016) Effect of surface roughness on the shaft resistance of non-displacement piles embedded in sand. Géotechnique 66(5):386–400
29. Tehrani FS, Arshad MI, Prezzi M, Salgado R (2018) Physical modeling of cone penetration in layered sand. J Geotech Geoenviron Eng 144(1):04017101
30. Tovar-Valencia RD, Galvis-Castro AC, Prezzi M, Salgado R (2018) Short-term setup of jacked piles in a calibration chamber. J Geotech Geoenviron Eng 144(12):04018092
31. Galvis-Castro AC, Tovar-Valencia RD, Salgado R, Prezzi M (2019) Effect of loading direction on the shaft resistance of jacked piles in dense sand. Géotechnique 69(1):16–28
32. Galvis-Castro AC, Tovar-Valencia RD, Salgado R, Prezzi M (2019) Compressive and tensile shaft resistance of nondisplacement piles in sand. J Geotech Geoenviron Eng 145(9):04019041
33. Ganju E, Han F, Prezzi M, Salgado R, Pereira JS (2020) Quantification of displacement and particle crushing around a penetrometer tip. Geosci Front 11(2):389–399
34. Tovar-Valencia RD, Galvis-Castro A, Salgado R, Prezzi M (2021) Effect of base geometry on the resistance of model piles in sand. J Geotech Geoenviron Eng 147(3):04020180
35. Ganju E, Galvis-Castro AC, Janabi F, Prezzi M, Salgado R (2021) Displacements, strains, and shear bands in deep and shallow penetration processes. J Geotech Geoenviron Eng 147(11):04021135
36. Tovar-Valencia RD, Galvis-Castro A, Salgado R, Prezzi M, Fridman D (2023) Experimental measurement of particle crushing around model piles jacked in a calibration chamber. Acta Geotech 18:1331–1351
37. Galvis-Castro AC, Tovar-Valencia RD, Prezzi M, Salgado R (2023) Effect of cyclic loading on the mobilization of unit base resistance of model piles jacked in sand. Acta Geotech 18:4747–4766

An Integrated Framework for Comprehensive Evaluation of Sustainability in Transportation Geotechnics

Krishna R. Reddy and Jagadeesh Kumar Janga

Abstract Transportation geotechnics is an important component in the construction of sustainable transportation infrastructure, playing a pivotal role in ensuring the stability, durability, and long-term performance of transportation networks. Incorporating sustainable practices into transportation geotechnics is essential for promoting environmental stewardship and achieving sustainable outcomes. While tools have been developed to evaluate the sustainability of engineering projects, many of them have primarily focused on environmental impacts, overlooking broader social and economic implications. To perform a comprehensive sustainability assessment, it is crucial to adopt a life cycle approach that considers multiple dimensions when evaluating a project's sustainability. In view of this, a triple-bottom line sustainability assessment framework called "Qualitative Assessment of Life Cycle Sustainability (QUALICS)" has been recently developed. This paper aims to provide a brief description of the QUALICS framework and demonstrate its application to assess the sustainability of design alternatives using three case studies of sustainable practices in pavements and transportation geotechnics: (1) fly ash and lime-stabilized sub-grades for pavement; (2) a streetscape with features such as recycled-aggregate base course, pervious-concrete sidewalks, and bio-retention islands; and (3) utilization of recycled-aggregate sub-base in pavements. Based on these case studies, this paper showcases the versatility and applicability of the QUALICS framework and highlights the significance of considering sustainability across various dimensions.

Keywords Sustainability metrics · Triple-bottom line · Recycled aggregates · Green infrastructure · Transportation geotechnics

K. R. Reddy (✉) · J. K. Janga
Department of Civil, Materials, and Environmental Engineering, University of Illinois Chicago, 842 West Taylor Street, Chicago, IL 60607, USA
e-mail: kreddy@uic.edu

B. Indraratna and C. Rujikiatkamjorn (eds.), *Recent Advances and Innovative Developments in Transportation Geotechnics*,
https://doi.org/10.1007/978-981-97-8245-1_3

1 Introduction

Transportation networks are an integral part of infrastructure that address the growing transportation needs, driven by population growth and increasing freight demands, ensuring that the evolving infrastructure can accommodate these challenges effectively. Transportation geotechnics is a sub-branch of transportation engineering which ensures the subsurface stability of transportation infrastructure. However, current practices in transportation geotechnics such as the use of materials like river sand and quarry-derived aggregates as the raw materials in base, sub-base, and subgrade courses of pavements are resource intensive and unsustainable in long-term. Hence, there is a growing need to replace these materials with more sustainable materials like recycled aggregate, slag, mine overburdens, and other appropriate industrial byproducts. Research has shown the potential of such materials to be employed as sustainable alternatives to traditional materials, used in base and sub-base courses [1, 2]. However, in order to understand the true sustainability of utilizing these sustainable materials in practice, it is important to quantify the same which can assist in the final decision-making to employ such materials.

The life cycle approach—which considers a project's entire life cycle from raw material processing to waste disposal—has gained traction in engineering design, with tools like life cycle assessment (LCA) evaluating environmental impacts across different stages of a project gaining prominence [3]. However, LCA predominantly addresses environmental concerns, overlooking social and economic dimensions encompassed in the triple-bottom line sustainability. The American Society of Civil Engineers (ASCE) has recently introduced a sustainability standard emphasizing on the importance of considering triple-bottom line sustainability in infrastructure development [4]. ASCE's standard for sustainable engineering focuses on considering project's entire life cycle, while also educating stakeholders about a project's net environmental, social, and economic impacts.

Evaluating environmental aspects involves metrics like global warming, ozone depletion, eutrophication, energy usage, fossil fuel depletion, and land use, often quantified through specialized tools. Economic impacts, on the other hand, encompass direct and indirect costs, the latter including social costs related to carbon emissions, and direct and indirect economic benefits. Cost–benefit analysis is a prevalent method for assessing the economic sustainability of a project. Further details on various tools available tools for economic sustainability assessment are presented in Reddy et al. [5]. As for social sustainability, gauging its impact is complex, lacking definitive metrics or tools. Yet, it remains a crucial facet within the triple-bottom line framework. To address this, a semi-quantitative tool for evaluating social sustainability is the Social Sustainability Evaluation Matrix (SSEM), developed by Reddy et al. [6]. In addition, tools such as the Environmental Justice Screen (EJ Screen), developed by the US Environmental Protection Agency (USEPA) [7], can also be used to gauge the impacts of various kinds of emissions on communities.

However, depending on the project type, the environmental, social, and economic sustainability analyses can consist of different kinds of variables (qualitative, quantitative, or semi-quantitative) that measure the sustainability of each of these aspects [8]. In addition, the calculated environmental sustainability can be higher for a given alternate solution, but it can be socially and/or economically less sustainable, making the decision-making process harder. Given the variability in the variable types and the possibility of ambiguity in decision-making, it is important to integrate the triple-bottom line aspects of sustainability in order to be able to properly gauge the sustainability of a given alternative and aid in making decisions. Current paper introduces a framework for this purpose, Quantitative Assessment of Life Cycle Sustainability (QUALICS) developed at the University of Illinois Chicago [8, 9]. This paper provides a brief description of QUALICS methodology and demonstrates the use of the same in transportation geotechnics and pavements' related projects with the help of three different case studies.

2 QUALICS Methodology

The QUALICS framework, which is used to perform an integrated triple-bottom line sustainability assessment, can be applied to any kind of engineering projects/product/activities. The stepwise methodology of the QUALICS framework is shown in Fig. 1, and the description of the same is provided below.

Step 1: A transportation engineering problem can have multiple viable solutions. For instance, the sub-base course of a flexible pavement can contain traditional quarry-derived aggregates or recycled aggregates derived from industrial byproducts, depending on whether the material is a technically sound alternative. To discern the most sustainable option between such alternatives, a comprehensive quantitative evaluation of the wider environmental, social, and economic consequences across all stages of the project's life cycle becomes crucial. Thus, the first step in the QUALICS framework encompasses the identification of potential design alternatives that can effectively serve the project's objective. This includes a thorough technical design for each alternative, aligned with the project's requisites and site specifications.

Step 2: This step involves identifying both qualitative and quantitative variables that closely correspond to the important environmental, social, and economic implications of various design alternatives. These variables are classified into different categories, encompassing requirements, criteria, and indicators. At the requirement level, variables contain the environmental, social, and economic aspects. Each requirement is subsequently divided into criteria, and the criteria are further broken down into indicators. The specific variables are tailored to the project's particulars, thus exhibiting variability across projects. Environmental criteria, for instance, comprise air and water usage, energy consumption, as well as impacts on land and ecosystems. Meanwhile, economic criteria encompass both direct costs (like materials, labor, equipment, transportation of materials, and end-of-life waste disposal) and indirect

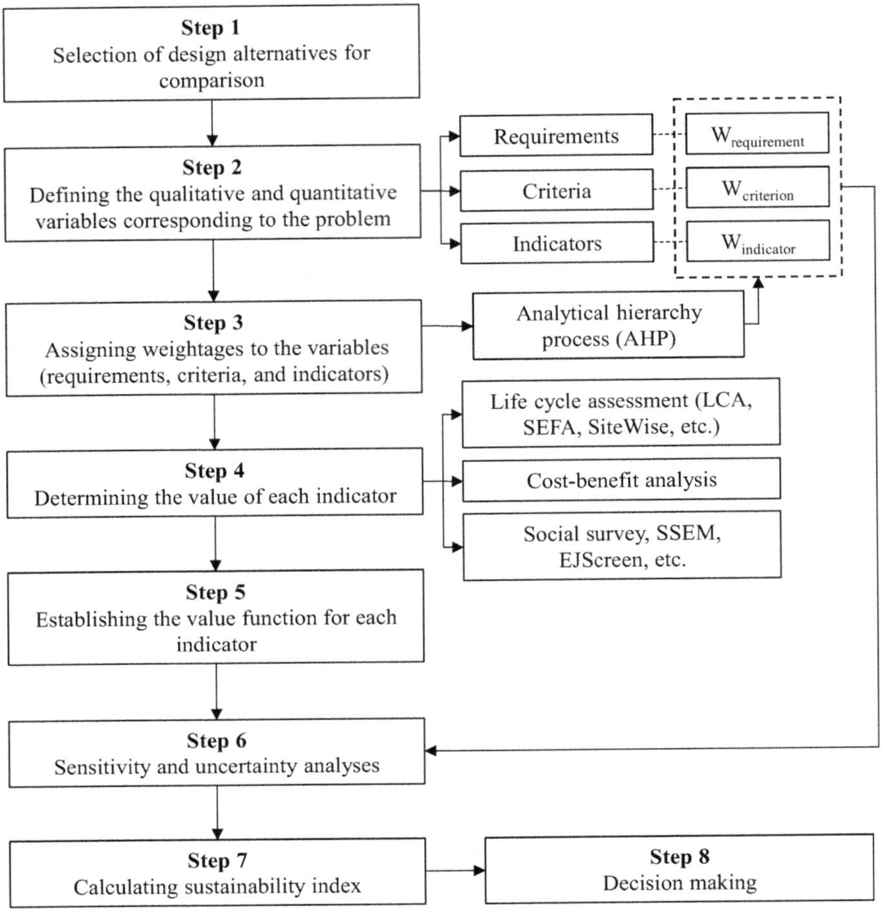

Fig. 1 Stepwise methodology for QUALICS framework

costs (social costs related to carbon emissions), in addition to direct and indirect benefits (revenue generation and employment creation among others). Social criteria span social-individual, socio-environmental, socio-economic, and socio-community impacts of a project.

The indicator-level variables signify impacts that trigger broader effects at higher level i.e., criteria level. In the QUALICS framework, environmental indicators can be quantified using various established tools for environmental impact assessment like LCA. Economic indicators encompass the direct costs tied to raw materials, labor, and equipment for construction, transportation, and end-of-life waste disposal as well as indirect costs such as the social costs of various emissions, in addition to the direct and indirect economic benefits, which can vary depending on the project. Social indicators also can differ for different projects and the design alternatives under consideration.

Step 3: This step includes attributing weightages to the variables within each category. This can be achieved following the analytic hierarchy process (AHP), a technique introduced by Saaty [10], or can be manually assigned depending on stakeholder preference and expert judgment. AHP is used to methodically guide judgments and allot priorities among different variables. The AHP methodology consists of several phases, starting with the problem definition and hierarchy structuring from requirement to indicator levels. Subsequently, a pairwise comparison matrix is formulated, facilitating the assessment of relative importance between different variables, which can then be used to assign weights to variables. Further insights into the application of AHP methodology within the QUALICS framework can be found in the work of Trentin et al. [9].

Step 4: In this step of the QUALICS framework, various indicators identified will be quantified. To find the values for environmental indicators, tools like LCA can be used, which looks at the whole process from obtaining raw materials to disposal of waste. For economic indicator values, both direct and indirect costs and benefits should be considered. For measuring social sustainability of the proposed alternatives for a given project, surveys can be conducted. Based on the inputs from experts, the survey can be designed and sent to stakeholders with varying degrees of relevance to the project. The results from the surveys help in understanding the importance of each social impact, and this information can be used to give final scores for each social impact indicator. Social impact assessments can also be performed using established frameworks like the SSEM framework [6].

Step 5: The indicators quantified in Step 4 generally have different metrics or units of measurement which makes it impossible to compare the impacts from different variables on the overall sustainability. Hence, to make the indicator values comparable across dimensions, the values of all the indicators have to be normalized. One commonly used normalization function, called the value function, can be used to normalize the indicators on a scale of 0–1 (0 indicating least sustainability/satisfaction, and 1 indicating the highest sustainability/satisfaction). The shape of the value function, as defined by Alarcon et al. [11], can vary depending the type of variable (linear/concave/convex/S-shaped). A detailed explanation on various forms of the value function was provided by Alarcon et al. [11].

Step 6: This step involves a sensitivity analysis in order to identify the variables that are outweighing the impacts of other variables. For example, if the transportation of raw materials is predominantly showing the highest impact among all the indicators, then a sensitivity analysis can be performed by varying the distances of transport to check the differences in overall environmental impacts due to such variations.

Step 7: In this step, the normalized values of different indicators will be integrated to obtain an overall sustainability index. The Integrated Value Model for Sustainability

Assessment (MIVES) methodology can be used to calculate the sustainability index [12]. In this methodology, the obtained indicator values on a scale of 0–1 (V_{ind}) are multiplied with their corresponding weightages (W_{ind}). The value of criteria (V_{cr}) can then be obtained by summing the products of indicator values to their respective weightages (Eq. 1). Similarly, the sum of the products of the values of the criteria with their corresponding weightages (W_{cr}) gives the values of requirement level variables (V_{req}) (Eq. 2). The final sustainability index (V_{final}) can then be obtained by summing the products of requirement values and their corresponding weightages (W_{req}) (Eq. 3).

$$V_{cr} = \sum_{1}^{n} V_{ind} \times W_{ind}, \quad (1)$$

$$V_{req} = \sum_{1}^{m} V_{cr} \times W_{cr}, \quad (2)$$

$$V_{final} = \sum_{1}^{k} V_{req} \times W_{req}, \quad (3)$$

where n, m, and k are the no. of variables in the indicator, criteria, and requirement categories, respectively.

Step 8: Final step of the framework, i.e., decision-making, involves deciding on the best alternative by comparing the sustainability index of all the alternatives. The alternative obtaining highest sustainability index value is considered the most sustainable option. Depending on the stakeholder preference of the environmental, social, and economic aspects, the overall sustainability index can vary. For example, if the stakeholders consider economic aspects as more important, then the weightage for the economic requirement can be increased accordingly. Overall, in this section, a brief description on the stepwise methodology of the QUALICS framework is provided. For a more detailed explanation regarding the QUALICS framework and implementation, refer to Trentin et al. [9] and Reddy et al. [8].

3 Case Studies

3.1 Case I—Sustainability Assessment of Fly Ash and Lime-Based Sub-grades for a Pavement

Reddy and Giri [13] compared the overall sustainability of two alternative materials: hydrated lime and Class C fly ash, utilized to stabilize the sub-grade containing glacial till under a newly constructed flexible pavement, in a rural location near Omaha,

Nebraska, USA. This investigation included the development of two pavement designs with two stabilization materials that results in enhanced bearing capacity of the sub-grade. Analysis of these designs demonstrated that the sub-grade stabilized with hydrated lime necessitated a thinner wearing surface and base in comparison to the sub-grade stabilized with Class C fly ash. A comprehensive sustainability assessment, encompassing all three dimensions of sustainability: environmental, social, and economic, was performed to compare the sustainability of both the materials.

To assess the environmental impacts, life cycle assessments (LCA) for both pavement design scenarios were conducted using SimaPro 8.1 software with Tool for Reduction and Assessment of Chemicals and Other Environmental Impacts (TRACI), 2.1 V1.01/US 2008. TRACI is a midpoint analysis method specific to the US conditions, developed by the USEPA. The environmental sustainability in TRACI is assessed using ten impact categories: global warming, ecotoxicity, eutrophication, carcinogenics, non-carcinogenics, fossil fuel depletion, ozone depletion, acidification, smog, and respiratory effects. Results from environmental assessment (Fig. 2a) showed that lime has relatively less environmental impacts compared to fly ash in most of the categories.

Economic analysis included direct costs—material, transport, and construction costs—and indirect costs—social cost of carbon. Results (Fig. 2b) indicated an almost similar direct and indirect costs of both the alternatives analyzed. To evaluate the

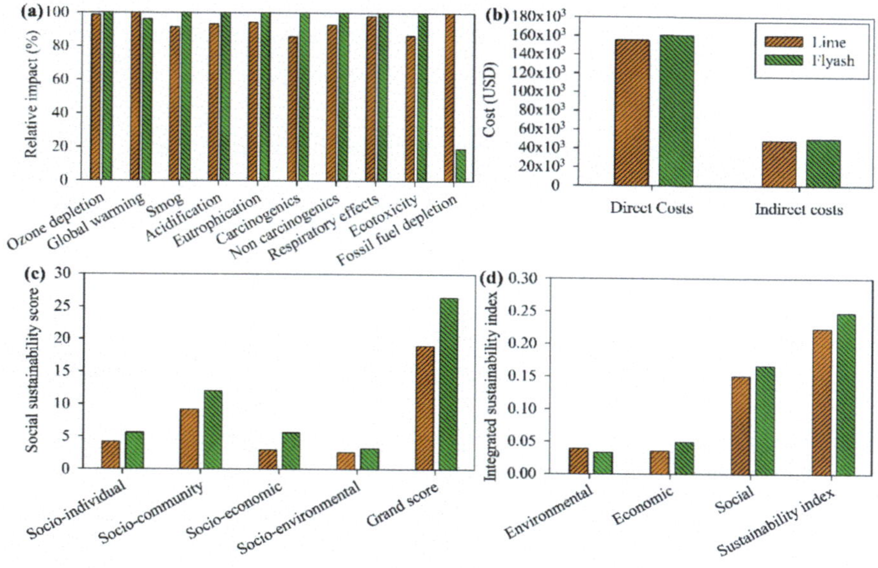

Fig. 2 Sustainability assessment of two different pavement systems: **a** environmental, **b** economic, **c** social, and **d** overall sustainability

social sustainability, ratings were given on various aspects defining the social impacts of the proposed materials (Fig. 2c). The results of the study were indecisive as the environmental impacts of lime were less than fly ash, and the economic impacts of both the cases were almost similar, whereas the social impacts showed better performance while using fly ash.

To address this, current study integrated the results from Reddy and Giri [13]—using the QUALICS framework—to obtain an integrated sustainability index. Results revealed that the use of fly ash is slightly more sustainable (Fig. 2d) compared to lime, owing to the significantly higher social benefits of using fly ash, while lime was only slightly more sustainable in the environmental and economic aspects.

3.2 *Case II—Sustainability Assessment of a Green Streetscape*

Reddy et al. [14] assessed various options for the reconstruction of a street located in downtown Oak Park, Illinois, USA. The conventional pavement design adheres to the Illinois Department of Transportation (IDOT) standards for both hot mix asphalt and concrete pavements. To pinpoint the most sustainable design, four distinct design options were compared in Reddy et al. [14], and the details of the alternatives compared are presented in Table 1. A comprehensive sustainability evaluation was undertaken to encompass the wide-ranging environmental, social, and economic aspects associated with these design alternatives.

Table 1 Different alternatives analyzed for the development of a street located in Oak Park, Chicago, IL, USA

Component	ATI	AGI	CTI	CGI
Pavement	Hot mix asphalt	Hot mix asphalt	Concrete	Concrete
Materials used in sub-layers	0% recycled	100% recycled	0% recycled	100% recycled
Curb and gutter	Concrete	Concrete	Concrete	Concrete
Islands	Conventional concrete	Bio-retention	Conventional concrete	Bio-retention
Sidewalks	Conventional concrete	Pervious concrete	Conventional concrete	Pervious concrete
Pavement marking	Standard reflectivity solvent-based	High reflectivity water-based	Standard reflectivity solvent-based	High reflectivity water-based
Street lights	High-pressure sodium vapor	LED	High-pressure sodium vapor	LED

Note Abbreviations: *ATI* asphalt pavement with traditional infrastructure, *AGI* asphalt pavement with green infrastructure, *CTI* concrete pavement with traditional infrastructure, *CGI* concrete pavement with green infrastructure

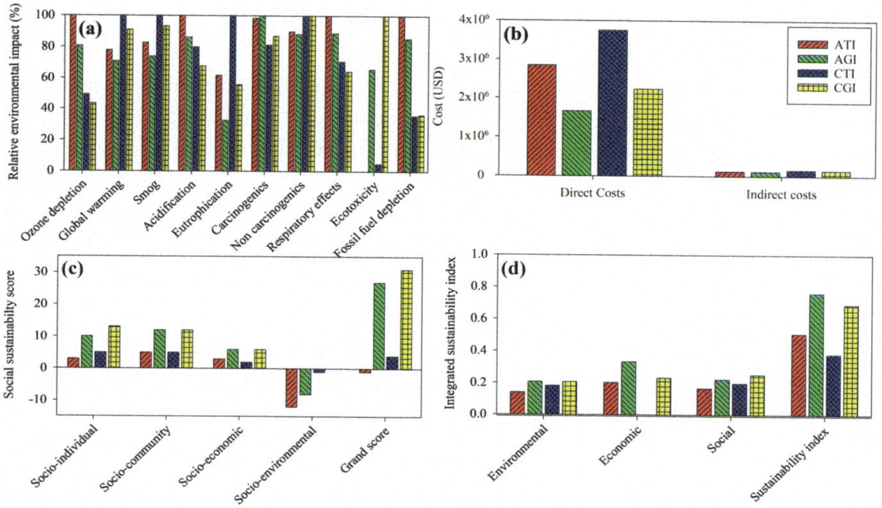

Fig. 3 Sustainability assessment of four different alternatives analyzed: **a** environmental, **b** economic, **c** social, and **d** overall sustainability

The assessment of sustainability involved multiple dimensions. Firstly, after the detailed technical design is done based on standard methods, an analysis was conducted to quantitatively measure the environmental impacts linked to each alternative. The results of environmental sustainability assessment are presented in Fig. 3a, indicating alternatives 1 and 3 with traditional infrastructure and virgin aggregates to have higher environmental impacts in most categories. Furthermore, a life cycle cost evaluation was carried out to account for the direct financial expenses tied to each design and the indirect costs associated with carbon emissions, i.e., social cost of carbon (Fig. 3b). However, the above-mentioned study did not perform a quantitative social impact assessment and an integrated triple-bottom line assessment.

In order to address these aspects, current study performed a social impact assessment using a rating-based framework called the SSEM framework developed by Reddy et al. [6]. The social impact assessment considered various aspects by dividing the social requirement into four criteria, namely: social-individual, social-economic, social-institutional, and social-environmental.

The social implications of using—asphalt and concrete pavements, sustainable elements like recycled aggregates for sub-base course, and green infrastructure for stormwater management—were considered. In addition, the enhancement of civic pride, improvement of quality of life due to well-paved streets and green infrastructure, and the socio-environmental benefits of using greener technologies like LED lights, pervious-concrete sidewalks, and bio-retention islands among others were also taken into consideration. Various indicators were identified in these lines, and each indicator was rated on a scale of − 2 to + 2 with − 2 representing the worst impact and + 2 being the score for ideal case. The ratings from these indicators were added to obtain a cumulative social sustainability score. The results obtained from the social impact assessment are presented in Fig. 3c.

Further, the environmental and economic sustainability indicators obtained from Reddy et al. [14] and the social sustainability indicators obtained from the SSEM were integrated to develop an integrated sustainability index using QUALICS framework. The results of the same are presented in Fig. 3d. Overall, it was found that the asphalt pavement with green infrastructure and other enhancements in terms of water-based reflectors and LED street lights fared the highest sustainability index considering all the triple-bottom line aspects.

3.3 Case III—Sustainability Assessment of Utilization of Recycled Aggregate in Pavement Design

Seyrfar et al. [15] performed a TBL sustainability assessment of a pavement using three different alternative options. The pavement under consideration was a 14-mile stretch of Interstate 90 highway, spanning from Schaumburg to Elgin in Illinois, USA. The Illinois Tollway Authority undertook the reconstruction of this pavement. The revamped road segment was built using jointed concrete pavement along with a sub-base composed of recycled aggregates. The previous pavement was removed and processed on-site to be utilized as part of the sub-base material. The aforementioned study compared the sustainability of this pavement to two other alternatives: one involving a similar pavement but a virgin aggregate sub-base and the other utilizing full-depth asphalt pavement with a similar recycled sub-base. To determine the most sustainable choice among these three pavement alternatives, the triple-bottom line sustainability framework, QUALICS, was employed.

The environmental impacts in this study were estimated with the help of SimaPro software using TRACI 2.1 V1.01/US 2008 impact assessment method. The economic sustainability was analyzed using a life cycle cost analysis including both direct and indirect costs. The indirect costs included the social cost of carbon emissions. The social sustainability aspect was evaluated by developing a public survey and utilizing the results to develop social sustainability score using Social Sustainability Evaluation Matrix (SSEM). Subsequently, the results of these three dimensions were integrated to compute a sustainability index using the MIVES methodology. The comprehensive findings ultimately indicated that the jointed concrete pavement with a recycled sub-base exhibited the highest sustainability index, as depicted in Fig. 4. Case II presented before indicated that asphalt pavements are more sustainable unlike this case. This is because this case considered the environmental and economic costs of long-term maintenance which is higher for asphalt pavements compared to concrete pavements.

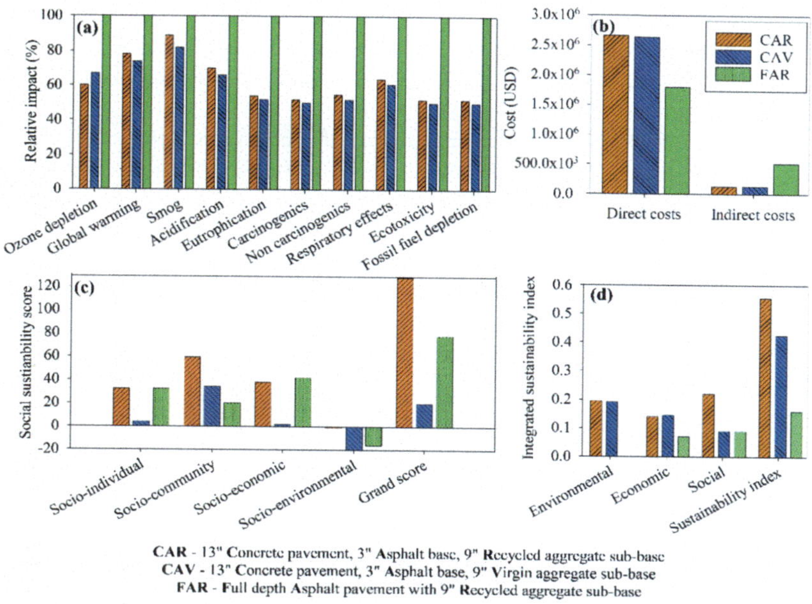

Fig. 4 Sustainability assessment of three different pavement systems: **a** environmental, **b** economic, **c** social, and **d** overall sustainability

4 Summary and Challenges

In light of global challenges such as a growing population and climate change, the importance of the concept of sustainable development is growing. The activities associated with transportation projects consume vast quantities of resources across the project's entire life cycle. Consequently, it becomes essential to quantify the broader environmental, social, and economic impacts stemming from such activities. This assessment is crucial for identifying the most efficient and sustainable course of action, thereby contributing to global sustainable development. However, most sustainability assessment tools often fail to consider the comprehensive social and economic impacts from a life cycle viewpoint, while integration of these triple bottom line aspects to judge the overall sustainability is a challenge in itself. In this regard, a novel quantitative sustainability assessment framework known as QUALICS has been introduced. This framework aims to facilitate the quantification of environmental, social, and economic impacts and the identification of the most sustainable design alternative. QUALICS leverages multi-criteria decision-making methods like MIVES and AHP, to quantify the sustainability of various options under consideration. This quantification is based on the relative significance and relevance of distinct criteria and indicators that represent the potential impacts arising from project activities on-site. The framework also relies on the expertise and judgment of professionals, thereby incorporating the innate complexity and reality of the system into the assessment. In this study, three case studies in the field of pavement geotechnics were presented demonstrating the applicability and flexibility of the QUALICS framework. These studies utilized the QUALICS framework to determine the most sustainable design alternative for implementation.

Future research should prioritize enhancing the economic dimension by integrating indirect costs and overall economic benefits that are often excluded. Moreover, the social sustainability assessment component within the QUALICS framework requires further enhancement. There's a need for a well-defined and properly structured social sustainability assessment tool, akin to the life cycle assessment methodology, with an ability to quantify and assess the social sustainability of different projects. Additionally, integrating the resilience aspect into sustainability assessment frameworks is crucial, given that resilience and sustainability are often interconnected.

References

1. Arulrajah A, Piratheepan J, Disfani MM, Bo MW (2013) Geotechnical and geoenvironmental properties of recycled construction and demolition materials in pavement subbase applications. J Mater Civ Eng 25(8):1077–1088
2. Mishra A, Das SK, Reddy KR (2023) Characterization and environmental sustainability of open pit coal mine overburden waste rock as pavement geomaterial. Transp Geotech 42:101094
3. Reddy KR, Cameselle C, Adams JA (2019) Sustainable engineering: drivers, metrics, tools, and applications. Wiley, Hoboken, NJ
4. ASCE (American Society of Civil Engineers) (2023) Standard practice for sustainable infrastructure. ASCE, Reston, VA
5. Reddy KR, Kumar G, Chetri KJ (2019) Green and sustainable site remediation: incorporating broader economic impacts. In: The 17th African regional conference on soil mechanics and geotechnical engineering
6. Reddy KR, Sadasivam BY, Adams JA (2014) Social sustainability evaluation matrix (SSEM) to quantify social aspects of sustainable remediation. In: ICSI 2014: creating infrastructure for a sustainable world. ASCE, Reston, VA, pp 831–841
7. USEPA (United States Environmental Protection Agency). EJScreen user guide version 2.1. https://ejscreen.epa.gov/mapper/help/ejscreen_help.pdf. Accessed 21 Nov 2023
8. Reddy KR, Kumar G, Chetri JK (2021) Quantitative assessment of life cycle sustainability (QUALICS): application to engineering projects. In: Indian geotechnical conference 2019: Geotechnics for INfrastructure Development & UrbaniSation (GeoINDUS). Springer, Singapore, pp 111–125
9. da S Trentin AW, Reddy KR, Kumar G, Chetri JK, Thomé A (2019) Quantitative assessment of life cycle sustainability (QUALICS): framework and its application to assess electrokinetic remediation. Chemosphere 230:92–106
10. Saaty TL (2008) Decision making with the analytic hierarchy process. Int J Serv Sci 1(1):83–98
11. Alarcon B, Aguado A, Manga R, Josa A (2010) A value function for assessing sustainability: application to industrial buildings. Sustainability 3(1):35–50
12. MIVES IV Project (2010) Quantification of sustainability in construction engineering with and without uncertainty. Ministerio de Ciencia y Education, Madrid, Spain
13. Reddy KR, Giri RK (2015) Assessing sustainability of ground improvement methods: quantitative triple bottom line framework and case study. In: Proceedings of international conference on infrastructure development for environmental conservation and sustenance (INDECS-15), Hosur, Tamil Nadu, India, pp 35–44

14. Reddy KR, Bakos W, Doubek B, Kumar G (2017) Sustainable streetscape: case of Lake Street in downtown Oak Park, Illinois, USA. In: ASCE India conference 2017. ASCE, Reston, VA, pp 610–618
15. Seyrfar A, Osman I, Yarveicy V, Akbuga M, Reddy KR (2021) Life cycle sustainability assessment of flexible pavement designs: a case study of Illinois Interstate Highway. In: Transportation research board 100th annual meeting, TRBAM-21-03544. Transportation Research Board

AI-Based Ballasted Track GPR Application and Development

Xuanyang Qin, Zhan Peng, and Guoqing Jing

Abstract The use of ground-penetrating radar (GPR) technology in railway maintenance and inspection has gained significant attention in recent years due to its non-destructive and efficient nature. This paper explores the application and development of artificial intelligence (AI) in conjunction with GPR for the assessment of ballasted tracks. The integration of AI algorithms enhances the interpretation of GPR data, providing a more accurate and automated method for detecting and predicting potential trackbed fouling. This article explores the innovative applications of machine learning and deep learning in GPR data processing for railway track engineering prospection, evaluating their potential, limitations, and future prospects, aiming to contribute to the advancement of railway infrastructure monitoring, with a focus on improving reliability, safety, and maintenance efficiency.

Keywords Ground-penetrating radar (GPR) · Railway ballasted track · Automated analysis · Artificial intelligence · Deep learning

1 Introduction

Railway transportation plays a pivotal role in global logistics, and the maintenance of railway tracks is crucial for ensuring safe and efficient operations. Traditional methods of track inspection often involve time-consuming and labor-intensive procedures. The track structures experience accelerated deterioration due to the high-speed train traffic, heavier axle loads, and increased volume, resulting in the need for

X. Qin · G. Jing (✉)
School of Civil Engineering, Beijing Jiaotong University, Beijing 100044, China
e-mail: gqjing@bjtu.edu.cn

Z. Peng
Infrastructure Inspection Research Institute, China Academy of Railway Sciences Co., Ltd., Beijing 100081, China

B. Indraratna and C. Rujikiatkamjorn (eds.), *Recent Advances and Innovative Developments in Transportation Geotechnics*,
https://doi.org/10.1007/978-981-97-8245-1_4

frequent maintenance interventions [1]. The emergence of ground-penetrating radar (GPR) has provided a non-destructive and rapid means of assessing the condition of railway ballasted tracks [2].

Currently, the hardware for GPR applications in railway infrastructure is well-established by various companies, but the software for processing and interpreting GPR data still requires advancements to meet the exponential growth in GPR data volume, driven by the substantial increase in inspection speed, data acquisition rate, and railway line mileage. Automated GPR data processing and interpretation techniques are crucial for enabling rapid railway track defect inspection, ultimately leading to real-time railway infrastructure health monitoring. This paper investigates the synergy between GPR technology and artificial intelligence (AI) for the comprehensive evaluation of ballasted tracks.

2 GPR Technology in Railway Track Inspection

GPR is a geophysical method that uses electromagnetic waves to image the subsurface. A pulse of electromagnetic energy is generated to the detected different dielectric properties using a radio wave source. Then the reflected energy is received and analyzed [3], as the measuring principle is shown in Fig. 1.

GPR data provides insights into the condition of railway tracks by capturing changes in reflection amplitude, arrival time of specific reflections, and signal attenuation. The processed GPR data is employed to assess the condition of ballast, sub-ballast, and subgrade layers, capturing the quality and thickness of track structures, such as voids, ballast degradation, water infiltration, and ballast fouling.

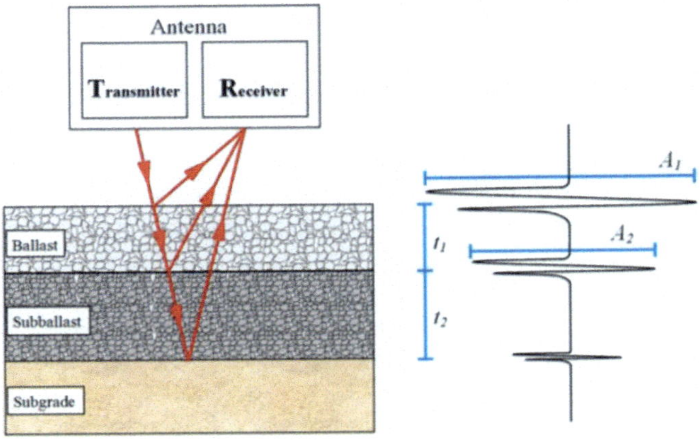

Fig. 1 Measuring principle of GPR, reproduced from [3]

The interpretation of GPR data for revealing the track defects is challenging for the following reasons.

- Large volume of data: With increased inspection speed and data acquisition rate, the volume of GPR data collected has grown exponentially. This can be overwhelming for data processing and interpretation.
- Signal complexity: Ground-penetrating radar (GPR) signals are often complex and challenging to interpret due to the presence of noise contamination, multiple reflections, and signal attenuation [4]. These factors can significantly hinder the ability to accurately assess the condition of railway infrastructure.
- Variability in track conditions: Railway tracks vary in design and construction, incorporating different types of ballast, subgrade materials, and track components. This diversity can lead to variations in GPR signal characteristics.
- Human interpretation limitations: The traditional interpretation of GPR data is often subjective, relying on the expertise and experience of the interpreter. This can lead to inconsistencies in defect detection and assessment, which is also laborious and time-consuming. This can lead to inconsistencies in defect detection and assessment.

Despite these challenges, GPR is a valuable tool for railway track inspection. Advancements in technology, such as machine learning (ML) and artificial intelligence (AI), offer promising solutions.

3 AI Integration for GPR Data Interpretation

The application of AI in GPR data interpretation involves the development of machine learning algorithms capable of recognizing patterns and anomalies within the radar data. Convolutional neural networks (CNNs) and other deep learning architectures are employed to automate the analysis process, reducing the dependency on manual interpretation. The integration of AI enhances the accuracy and efficiency of track inspection, enabling real-time decision-making.

As a non-destructive testing (NDT) method for inspection, GPR has been used for various infrastructures such as road pavements [5], tunnel linings [6], and archaeological prospection [7]. Automatic GPR signal recognition using ML methods can be divided into three main phases: data preprocessing, feature extraction, and defect classification.

3.1 Data Preprocessing

GPR measurements often employ wideband signals to capture multiple interfering signals simultaneously, leading to a more comprehensive representation of reflected wave characteristics. However, to effectively extract defect information using AI models, it is crucial to eliminate these interfering signals, which are basically inherent interference originating from components within the GPR system itself, interference from external sources such as radio and communication signals, and ground or underground/subterranean interference.

Liu et al. [8] used automatic gain control (AGC) and autoregressive sliding average spectral estimation (ARMA) methods to increase the signal-to-noise ratio (SNR). Then the variations of dual trip time (or depth) and mileage are reflected in the GPR image features. These image features are the reflected wave amplitude, waveform, phase, and spectrum of the inter-layer interface of the subgrade.

Generative adversarial networks (GAN) can serve as data-driven clutter removal methods in GPR image processing. Wang et al. [9] developed a GAN-based rebar clutter elimination network for void detection. Similarly, Ni et al. [10] employed a Declutter-GAN to transform cluttered GPR images into clear B-scan data. GANs can also be leveraged for data augmentation, enabling the generation of arguably pseudo-realistic-looking images. Brockner et al. [11] explored the potential of GANs to expand the training dataset for GPR applications, addressing the challenges of imbalanced and limited labeled data. Qin et al. [12] proposed a novel GAN framework incorporating convolutional layers to synthesize B-scan images. While GANs offer an effective means of dataset expansion and automated data processing, their convergence can be challenging due to the risk of mode oscillatory or collapse.

3.2 Feature Extraction

The extracted features of A-scan data should possess three essential properties: (1) invariance to time shifts and time scales, ensuring robustness against variations in signal timing and duration; (2) insensitivity to noise and multiple reflections, enhancing the ability to distinguish between true defects and spurious reflections caused by noise or complex subsurface structures; and (3) ease of discrimination, enabling effective classification of different types of defects [13].

In contrast to single-channel features that focus on the intensity values of GPR images, B-scan features capture more comprehensive visual characteristics, encompassing color, texture, and shape. Shape features directly describe the morphological attributes of track defects, such as their size, elongation, and curvature. However, the relationship between color and texture features and specific track defect types is often implicit and complex, requiring extensive training with multiple classifiers to effectively differentiate between different defect classes.

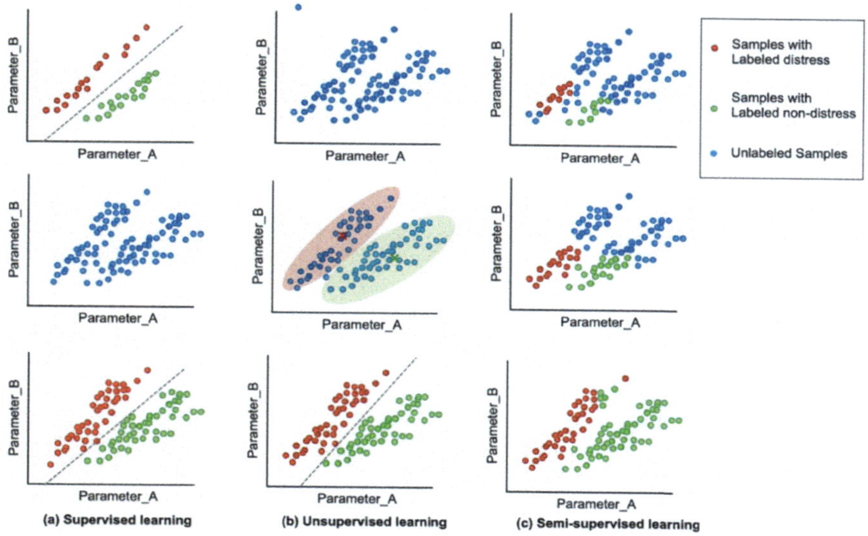

Fig. 2 Machine learning methods for A-scan-based identification, reproduced from [14]

3.3 Defect Classification

Track defect classification algorithms include unsupervised, supervised, and semi-supervised machine learning approaches. Unsupervised and supervised learning methods are the most widely employed techniques in the realm of automatic recognition of A-scan data [14], as shown in Fig. 2.

Deep learning has revolutionized various computer vision tasks, including image classification, object detection, and semantic segmentation. Among deep learning architectures, convolutional neural networks (CNNs) and recurrent neural networks (RNNs) have emerged as frontrunners for GPR defect identification [15]. Several advanced CNN architectures, including AlexNet, convolutional support vector machine (CSVM), Region CNN (R-CNN), you only look once (YOLO), and single shot multibox detector (SSD), have been successfully employed for RSD detection from B-scan images. Recurrent neural networks (RNNs) and generative adversarial networks (GANs) have also demonstrated promising results in this domain.

Under the challenge of limited labeled GPR data for defect detection, Kim et al. [16] proposed a transfer learning approach using the AlexNet architecture. As shown in Fig. 3, they leveraged the pre-trained weights of AlexNet, originally trained on a massive dataset of natural images, to effectively classify defects from B-scan images with minimal training data. This transfer learning strategy significantly reduced the training time and improved the classification accuracy compared to training the model from scratch.

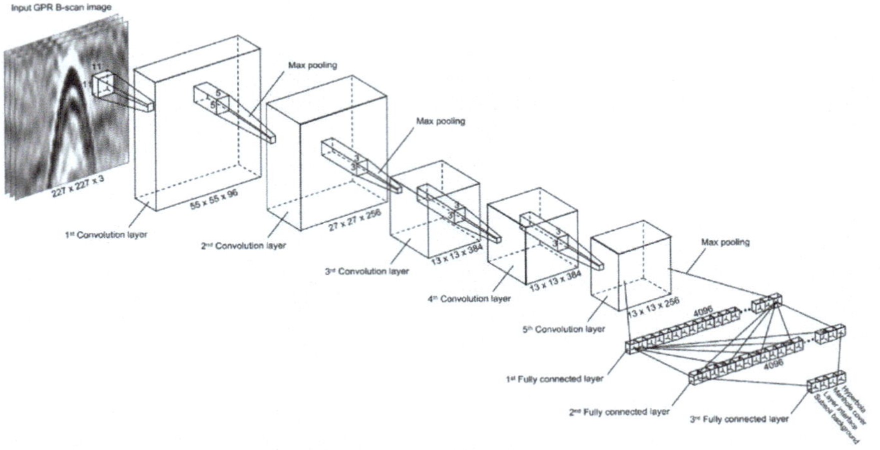

Fig. 3 Architecture of CNN used for B-scan image classification [16]

A-scan-based defect detection retains all applicable information without signal transformation losses. The method relies on extracting and classifying features from A-scan signals in the time, frequency, and time–frequency domains. However, A-scan-based algorithms suffer from three key limitations: (1) unstable feature extraction in noisy and heterogeneous environments, (2) heavy reliance on manual feature extraction, and (3) poor representation of defects spanning multiple channels.

Deep learning has shown promise for GPR B-scan image processing, but its generalizability and accuracy are often limited by irregular target shapes, obscuring backgrounds, and insufficient or unbalanced training data. Two main approaches address these challenges: (1) novel deep learning frameworks for small-sample training (few-shot and zero-shot learning) and (2) incorporation of domain knowledge into the deep learning pipeline (meta-learning and curriculum learning).

4 GPR-Based Intelligent Data Analysis and Track Assessment Approaches

The algorithms employed in railway track defect detection reflect the complexity of identifying and classifying various defect types. Unsupervised learning models excel at detecting anomalies and abnormal subsurface conditions without the need for labeled data [17]. Morphological methods are particularly effective in pinpointing specific forms of defects, such as cracks and voids. Supervised learning approaches like support vector machines (SVMs) and random forests are adept at distinguishing between two distinct categories of defects: voids and water-rich distress, owing to the significant differences in dielectric constants between water, air, and trackbed materials.

At present, limited studies have been found on correlating GPR-based data (ballast fouling) with track assessment (track geometry defects or deterioration rates). Wang et al. [18] estimated the occurrence of track geometry defects using substructure data measured by GPR, rail seat abrasion data, traffic data, and multiple parameters. Recursive feature elimination (RFE) was subsequently applied to reduce data dimensionality. Feature extraction techniques were employed to generate input variables for the machine learning models. A maintenance model was proposed.

Early work by [19] explored the connection between ballast layer performance and track geometry defects. Few studies have ventured to predict the precise relationship between track geometry defect development and ballast fouling. Goodarzi et al. [20] predicted Yellow-tag Defect in the Next Inspection (YDNI) using logistic regression and gradient boosting machine (GBM). GPR data were used as additional explanatory variables. A combination of four conditions is employed to identify maintenance interventions. In the k-fold cross-validation method, the entire dataset is randomly shuffled and then partitioned into k equal-sized folds, as shown in Fig. 4. The maintenance interventions can be explained in Fig. 5.

Deep learning methods stand out for their versatility in detecting all types of railway track defects. They identify defects and provide approximate estimations of their physical sizes. However, deep learning algorithms are susceptible to overfitting and underfitting due to limitations in the GPR dataset, potentially affecting the generalizability of their results. Within the realm of deep learning techniques, the YOLO network has emerged as the most widely used pre-trained railway track defects recognition model. Meanwhile, Mask R-CNN has established itself as the most prevalent semantic segmentation model for railway track defect detection. However, its application is currently restricted to small-scale and specific railway track defect types due to the scarcity of training samples.

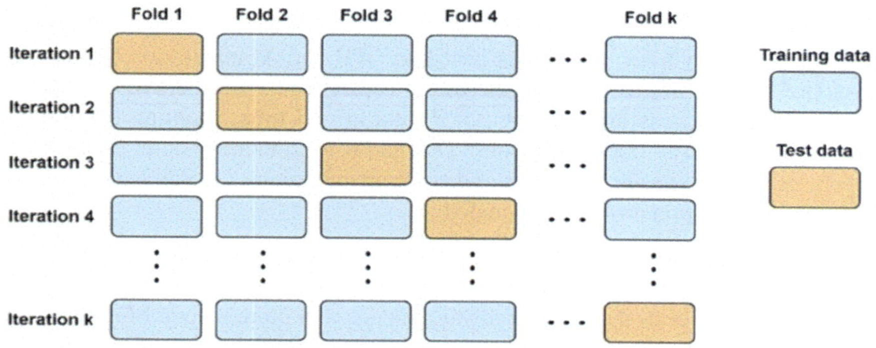

Fig. 4 Train-test split in k-fold cross-validation method [20]

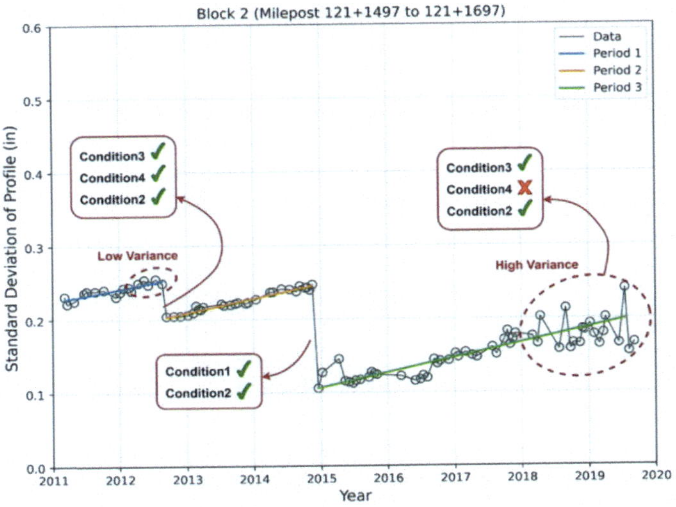

Fig. 5 Performance of automatic maintenance detection on two blocks [20]

The research on the correlation between GPR data and track defects is still in its early stages. More research is needed to develop more accurate and reliable methods for using GPR data to detect and predict track defects. ML models can be trained on large datasets of GPR data and track defect data to learn the patterns that are indicative of different types of defects. This information can then be used to develop new GPR-based inspection tools that can automatically detect and classify track defects.

5 Challenges and Future Directions

Despite the railway track assessment based on GPR and AI integration has not yet developed into a complete system, the future is promising. AI and ML algorithms can be used to process and interpret GPR data in real time, enabling continuous monitoring of railway track conditions. This can help to identify defects early and prevent them from causing accidents or disrupting operations. Overall benefits of AI for GPR data interpretation are concluded:

- Increased Accuracy. AI and ML algorithms can achieve higher accuracy in defect detection and classification compared to manual interpretation.
- Reduced Cost. With the higher railway mileage and more demand for ballasted track inspection. AI- and ML-based systems can automate the defect detection process, reducing the cost of labor and equipment.

- Improved Safety. AI and ML can help to identify defects early, which can prevent accidents and improve safety.

For future prospects, integrating AI and ML in the revolutionizing of GPR data presents several challenges that need to be addressed.

- Overcoming Noise Contamination. AI and ML algorithms must be robust enough to identify and remove this noise, including background clutter, interference from nearby objects, etc.
- Tackling Multiple Reflections. GPR signals can undergo multiple reflections within the subsurface, making it difficult to distinguish between true reflections and spurious echoes. AI and ML techniques need to be capable of disentangling these complex reflections to accurately interpret the GPR data.
- Mitigating Signal Attenuation. Algorithms should be able to compensate for signal attenuation to effectively detect defects at various depths.
- Automating Defect Segmentation. AI and ML techniques are needed to automate segmentation, enabling efficient defect detection and reducing the reliance on manual interpretation.

In developing new data processing methods, more research should be conducted on the following topics.

- The development of new ML models for detecting and predicting track defects.
- The collection of more GPR data and track defect data.
- The validation of GPR-based inspection tools in real-world settings.

Besides the design of ML algorithms, standardization, regulations, and collaboration with the railway industry are also important.

- Data Availability. Training and validating ML models require large amounts of high-quality GPR data, which can be challenging to collect and annotate.
- Generalizability. Intelligent approaches need to be adaptable to different track conditions, environments, and GPR equipment to ensure generalizability across various railway networks.
- Explainability. The interpretability of ML models is crucial for gaining trust and understanding the decision-making process behind defect detection and classification.
- Integration with Railway Operations. Intelligent GPR-based track assessment systems need to be seamlessly integrated into existing railway infrastructure and maintenance workflows for effective implementation.

By addressing these challenges and pursuing these promising research directions, AI and ML have the potential to revolutionize railway track inspection using GPR, leading to safer, more reliable, and more efficient railway operations worldwide.

6 Conclusion

The integration of AI with GPR technology holds great promise for advancing the field of railway ballasted track inspection. AI and ML have the potential to significantly improve the interpretation of GPR data for railway track inspection, leading to increased accuracy, reduced cost, and improved safety. This paper has presented an overview of the synergistic relationship between GPR and AI, emphasizing the benefits of enhanced accuracy and efficiency in ballasted track assessment. By combining GPR data with AI and ML techniques, these approaches can automate defect detection, improve accuracy and consistency, and enable continuous monitoring of track conditions. As research in this area continues, intelligent GPR-based track assessment is poised to revolutionize railway maintenance practices, ensuring the safety and reliability of railway transportation systems.

References

1. Fontul S, Paixão A, Solla M, Pajewski L (2018) Railway track condition assessment at network level by frequency domain analysis of GPR data. Remote Sens (Basel) 10(4). https://doi.org/10.3390/rs10040559
2. Guo Y, Liu G, Jing G, Qu J, Wang S, Qiang W (2022) Ballast fouling inspection and quantification with ground penetrating radar (GPR). Int J Rail Transp 1–18. https://doi.org/10.1080/23248378.2022.2064346
3. Wang H, Silvast M, Markine V, Wiljanen B (2017) Analysis of the dynamic wheel loads in railway transition zones considering the moisture condition of the ballast and subballast. Appl Sci (Switzerland) 7(12). https://doi.org/10.3390/app7121208
4. Oskooi B, Parnow S, Smirnov M, Varfinezhad R, Yari M (2018) Attenuation of random noise in GPR data by image processing. Arab J Geosci 11(21). https://doi.org/10.1007/s12517-018-4035-z
5. Rasol M et al (2022) GPR monitoring for road transport infrastructure: a systematic review and machine learning insights. Constr Build Mater 324. https://doi.org/10.1016/j.conbuildmat.2022.126686
6. Yue Y et al (2024) Automatic recognition of defects behind railway tunnel linings in GPR images using transfer learning. Measurement 224:113903. https://doi.org/10.1016/j.measurement.2023.113903
7. Küçükdemirci M, Sarris A (2022) GPR data processing and interpretation based on artificial intelligence approaches: future perspectives for archaeological prospection. Remote Sens 14(14). https://doi.org/10.3390/rs14143377
8. Liu H et al (2023) Combined CNN and RNN neural networks for GPR detection of railway subgrade diseases. Sensors 23(12). https://doi.org/10.3390/s23125383
9. Wang Y et al (2022) RCE-GAN: a rebar clutter elimination network to improve tunnel lining void detection from GPR images. Remote Sens (Basel) 14(2). https://doi.org/10.3390/rs14020251
10. Ni ZK, Shi C, Pan J, Zheng Z, Ye S, Fang G (2022) Declutter-GAN: GPR B-scan data clutter removal using conditional generative adversarial nets. IEEE Geosci Remote Sens Lett 19. https://doi.org/10.1109/LGRS.2022.3159788
11. Brockner B, Dowdy JL, Anderson DT, Veal C, Scott GJ, Ball JE (2018) Generative adversarial networks for ground penetrating radar in hand held explosive hazard detection, vol 10628, pp 306–323. https://doi.org/10.1117/12.2307261

12. Qin H, Zhang D, Tang Y, Wang Y (2021) Automatic recognition of tunnel lining elements from GPR images using deep convolutional networks with data augmentation. Autom Constr 130. https://doi.org/10.1016/j.autcon.2021.103830
13. Dogan M, Turhan-Sayan G (2016) Preprocessing of A-scan GPR data based on energy features, vol 9823, pp 415–422. https://doi.org/10.1117/12.2223946
14. Liu C, Du Y, Yue G, Li Y, Wu D, Li F (2024) Advances in automatic identification of road subsurface distress using ground penetrating radar: state of the art and future trends. Autom Constr 158:105185. https://doi.org/10.1016/j.autcon.2023.105185
15. Wang S, Liu G, Jing G, Feng Q, Liu H, Guo Y (2022) State-of-the-art review of ground penetrating radar (GPR) applications for railway ballast inspection. Sensors 22(7):2450. https://doi.org/10.3390/s22072450
16. Kim N, Kim K, An YK, Lee HJ, Lee JJ (2020) Deep learning-based underground object detection for urban road pavement. Int J Pavement Eng 21(13):1638–1650. https://doi.org/10.1080/10298436.2018.1559317
17. Sresakoolchai J, Kaewunruen S (2022) Railway defect detection based on track geometry using supervised and unsupervised machine learning. Struct Health Monit 21(4):1757–1767. https://doi.org/10.1177/14759217211044492
18. Wang X, Liu X, Euston TL (2023) Relationship between track geometry defect occurrence and substructure condition: a case study on one passenger railroad in the United States. Constr Build Mater 365. https://doi.org/10.1016/j.conbuildmat.2022.130066
19. Basye C, Wilk S, Gao Y. Ground penetrating radar (GPR) technology evaluation and implementation
20. Goodarzi S, Kashani HF, Oke J, Ho CL (2022) Data-driven methods to predict track degradation: a case study. Constr Build Mater 344. https://doi.org/10.1016/j.conbuildmat.2022.128166

The Importance of Rock Mass Characterization for Rock Socketed Foundations

Vasantha Wijeyakulasuriya, Siva Sivakumar, and Ajith Dissanayake

Abstract Rock mechanics investigations undertaken by the Department of Transport and Main Roads (TMR) in Queensland, Australia, as part of the pre-tender information for the Second Gateway Bridge (Duplication) have been revisited. In addition to the conventional rock substance characterization, this investigation included rock direct shear tests (CNS and CNL) and a significant number of rock pressuremeter testing to provide specialist rock mechanics information for the design of rock socketed foundations. This paper is focused on the major rock lithology intercepted, i.e., sandstone. Sandstone in the Sydney basin has been widely researched, and the findings from this study are compared with the published information on the Sydney Sandstone as well as from other sources. Findings published by the Second Gateway Bridge Designers are used for comparison where relevant, especially on the test piles which can serve as a useful reference.

Keywords Rock sockets · Shaft resistance · Rock characterization

1 Introduction

For major infrastructure projects undertaken by State Road Authorities, it has been common to undertake reasonable pre-tender geotechnical investigations, as the need to understand the geotechnical risk profile, ranks very high in major project risk assessments. In addition to informing the owner of the asset regarding the geotechnical risk involved, it assists the owner to decide on the procurement strategy regarding risk sharing. Prior geotechnical investigations enable proponents to evaluate the risks in their tender submissions, and this may be reflected in the pricing submissions. This paper is based on the pre-tender rock mechanics investigations

V. Wijeyakulasuriya (✉)
Tonkin+Taylor, Spring Hill, QLD, Australia
e-mail: VWijeyakulasuriya@tonkintaylor.com.au

S. Sivakumar · A. Dissanayake
Department of Transport and Main Roads (TMR), Brisbane, QLD, Australia

B. Indraratna and C. Rujikiatkamjorn (eds.), *Recent Advances and Innovative Developments in Transportation Geotechnics*,
https://doi.org/10.1007/978-981-97-8245-1_5

undertaken by TMR (Transport and Main Roads, Queensland, Australia) for the existing Second Gateway Bridge (Duplication), now known as the Sir Leo Hielscher Bridge, which is about 50 m downstream from the First Bridge. This paper has re-assessed the rock mechanics database and has identified those aspects that may be of wider interest to the geotechnical community at large.

The paper focusses on geotechnical rock substance/rock mass property characterization learnings identified from the review of the data. This data constitutes part of the geotechnical data accumulated for the 260 m central river-span foundations. The focus on rock characterization bears heavily on rock socketed foundation aspects and has been viewed against recent findings from published research.

2 Foundation Geology

2.1 Site Geology

The findings reported in this section on site geology and engineering geology of rock mass conditions are broadly in agreement with the previously reported literature, by Day et al. [1], on the Second Bridge, and Williams [2] on the First Bridge. The central span foundations of both bridges are founded within the Aspley-Tingalpa Formation. This formation was formed during the late Triassic Age from terrestrial alluvial sediments including carbonaceous material (low-grade coal). These sediments were deposited over Brisbane Tuff and Neranleigh-Fernvale Beds [3].

At the bridge sites, the geological formation consists of sandstone and mudstone with subordinate siltstone, claystone, carbonaceous shale, and low-grade coal seams, underlying the estuarine/alluvium deposits, within the investigated depths [1, 2]. Strata of Aspley-Tingalpa Formation are seen on exposures as sub-horizontally northeast dipping at approximately 10° in the project locale.

The study is focused on rock conditions encountered below the low-grade coal measures. Only rock core data from the investigated boreholes located within the central pier section of the Second Bridge have been taken into consideration for this study. Five boreholes under each central span river pier were advanced for this preliminary phase. At the southern river pier, the borehole depths ranged from 25.2 to 37.2 m, with the underside of the deepest coal intercepted including the overlying sediments, ranging from 12.4 to 19.4 m depth. The corresponding data at the northern pier ranged from 60.3 to 67.9 m for borehole depths and for the coal layer from 34.1 to 41.2 m. Based on an approximate 200 m of rock core retrieved (below the coal beds), the percentage of sandstone/sandstone dominant lithologies was found to be about 60%.

2.2 Engineering Geology of Rock Mass Conditions

For this study, the lithologies of Aspley-Tingalpa Formation have been taken into consideration. The intercepted rock types are classified into two major categories viz., sandstone (includes sandstone dominant interbedded rocks) and the other comprises fine grained sedimentary rocks (mudstone, siltstone and claystone and their interbedded derivations) for the purposes of this study. This study is limited in scope to an engineering assessment of the major lithology, i.e., sandstone/sandstone dominant rocks.

Based on our experience with sandstone formations across Queensland, geomechanical properties, quality, and durability could vary markedly from poorly cemented, very low strength to well cemented, durable, low to medium strength, and more competent rock masses. Therefore, characterization of sandstones in Queensland is a relatively challenging exercise and available literature on engineering characterization of these sandstone lithologies is very limited and mostly confined to some formations encountered in Southeast Queensland.

2.3 In-Situ Stress Field

As a result of past tectonic events, the major principal stress in the ground now has been measured either in the horizontal plane, or in the vertical plane, i.e., the gravitational stress. Major geological erosional effects have also been recorded which invariably leave a higher horizontal stress in the ground. Understanding the natural in-situ stress field (i.e., direction and magnitude) within the foundation rock mass is an important parameter to predict the possible stability issues not only for large and deep rock socket excavations but also for rock caverns and tunneling in rock. These in-situ stress field effects are likely to be more relevant at deeper depths, Pells [4]. Pells [4] citing the work of others has reported some work carried out in the Sydney basin and has indicated the *major principal stress* in the ground to lie in the *horizontal plane*. Rock pressuremeter testing undertaken as part of the reported investigation and discussed in Sect. 5 will address some of the potential implications due to the natural in-situ stress field.

Signs of the post-Triassic tectonic movements were noted during the initial site investigation on the exposed rock mass on existing cut slopes and foundation excavations in and around the southern approach area of the original bridge. North-west striking, steeply dipping, normal faults were observed within the exposed rock mass of the Aspley-Tingalpa Formation (Fig. 1 after [1]). Normal faulting or crustal extensional tectonics has been shown to be associated with in-situ stress fields where the *vertical gravitational stress* is the *major stress*, e.g., [5]. The observed evidence may not be adequate to make a definitive assessment of the in-situ stress field in the region.

Fig. 1 Normal fault, interpreted after [1]

In addition, the presence of frequent subvertical brecciated and shear zones with thickness up to 1700 mm was observed in the rock cores that were recovered from the deeper mudstone beds. Implications of such features for design and construction have been addressed by the foundation designers [1].

3 Overview of Rock Mass Characterization Requirements for Rock Socket Design

Field observations on instrumented rock sockets have shown that shaft resistance is generally mobilized at significantly smaller displacements than end-bearing resistance. This is understandable as the two load transfer mechanisms are quite different. Shaft resistance is due to frictional sliding along the concrete-rock interfaces with or without asperities. Frictional sliding over asperities of competent rock strength has a positive impact on both the ductility of shear response as well as contributing to enhance side-shear resistance aided by stiffness support from the surrounding rock mass by limiting/suppressing radial cracking. Comparatively, this mode of resistance has a higher stiffness, but less ductility of shear response. On the other hand, a capacity limiting state in end-bearing resistance is rarely reached and the common scenario is increasing resistance with increasing vertical displacement of the pile toe. This mode of resistance is less stiff within the same rock type but more ductile in response. Since the serviceability requirement of a structure is commonly dictated by settlements, in situations where the dominant loading is axial compression, the full base resistance is unable to be exploited.

In view of the above considerations, rock socketed piles generally carry a predominant portion of the working/serviceability load in side-shear resistance even if the rock socketed piles are designed as complete piles to exploit both side-shear resistance and end-bearing. For various reasons, socketed piles are sometimes designed to resist the loads entirely in side-shear, e.g., lack of confidence in ensuring reasonable cleanliness of the socket bases. For similar reasons, socketed piles are commonly designed to carry at least the serviceability load in side-shear resistance. For tension piles or piles in a pile group which is the more common scenario, that are subjected to uplift, for example, under lateral loading conditions, it will be only the side-shear resistance that is mobilized and therefore able to be exploited.

In routine design, it is more common for socketed piles to be designed to ensure adequate load capacity with subsequent checks on settlement. This translates into side-shear resistance socketed piles as well. However, to carry out a rational design of an individual socket, it is essential to have a tolerable settlement criterion for the pile group. This is a much wider and a complex soil/rock-foundation-superstructure interaction problem involving significant structural engineering input or largely dictated by them. Therefore, although, in general, side-shear-only socketed piles can be controlled by either load or settlement, the combined structure interaction demands on group effect may dictate that individual sockets be controlled by settlement.

Given the complex issues involved, for significant projects, one needs to go well beyond the rock substance characterization paradigm. The need for rock mass characterization is overwhelming. In this context, CNS/CNL concrete-rock joint interface testing, and rock pressuremeter testing to characterize rock mass stiffness, were carried out. Instrumented, Osterberg-based test piles formed an integral component, in the final comprehensive design/during construction investigation by others [1].

4 Intact Rock (Substance) Characterization

Most rock mechanics design methods addressing stability and stress-deformation-related issues such as foundations and tunneling are centered around, as initial inputs, on the intact or the substance properties of rocks, i.e., the uniaxial compressive strength, σ_{ci} (UCS) and Young's modulus, E_i. For example, one of the important components of the Hoek–Brown system for rock masses is to reduce laboratory-based material inputs into appropriate in-situ values. Therefore, intact rock characterization takes a dominant role in contemporary rock engineering practice.

4.1 Descriptive Statistics of the Sandstone Database

Substance properties of the sandstone: Table 1 details the σ_{ci} database in terms of Pells [4], five-tier intact strength classification. This classification has three attributes, i.e., σ_{ci}, defect spacing, and allowable seams. Only the σ_{ci} attribute is listed in Table 1. Since in the context of rock socket interface roughness, high strength rocks produce less roughness due to the methods required for high strength rock excavation [6], we have introduced Class "H" to capture the next higher strength category. See also the histogram shown in Fig. 4. According to the strength classification by Pells [4], the subject intact strengths, σ_{ci}, can be classified as low to medium strength. The rightmost column in Table 1 lists some substance property values. One striking observation is the reported E_i values (range 0.7–7.6) are much lower than the corresponding reported range for the Sydney sandstone (8–14). In contrast, this study has found the intact strength values to be stronger than reported in [4]. It should be noted that the property values given in Table 1 are dataset/site-specific. No measures were taken to saturate the samples.

To provide some contextuality, according to a global database on intact rock properties, ROCK/9/4069 [7], the *mean* COV values for sedimentary intact rocks are as follows (relevant Table 1 values are shown below in parentheses, for comparison):

Unit weight = 0.05 (0.03)
σ_{ci} = 0.32 (0.46)
E_i = 0.4 (0.45).

However, the reported results in Table 1 are within the ranges for COV reported in [7]. This global database is compiled from 184 studies and consists of 4069 intact rock data points, and about 60% is on sedimentary rocks.

Table 1 Classification of sandstone[a]

Class	σ_{ci} (MPa)	% of σ_{ci} test results ($n = 60$)	Property, mean, range; COV (= mean/std. dev.), no. of samples; [Pells, 2004, Class I/II]
H	> 48	15	Unit weight (kN/m^3): 24.5, 23–26; 0.03, 57; [24]
I	> 24	77	σ_{ci} (MPa): 35, 9.7–130; 0.46, 60; [12–50]
II	> 12	93	E_i (GPa): 3.4, 0.7–7.6; 0.45, 52; [8–14]
III	> 7	100	MR (E/σ_{ci}): 105, 47–209; 0.32, 52
IV	> 2	100	$I_{s50}D$ (MPa): 0.8, 0.2–1.9; 0.47, 51
V	> 1	100	$I_{s50}A$ (MPa): 1.5, 0.4–3.4; 0.44, 51

[a] SW weathering grade, 51–52 mm average sample size (diameter); rock testing carried out in laboratories at TMR, Monash University and University of Queensland

Transformation models (deterministic/point estimates): Figure 2 shows the transformation models between σ_{ci} versus I_{s50} for both diametral and axial point load values, and Fig. 3, E_i versus σ_{ci}. The linear regression lines have been forced to pass through the origin. According to [7], however, the most effective transformation model for E_i is based on V_p, i.e., P-wave velocity measurements on intact samples.

Strength anisotropy and implications: The ratio of axial point load mean value to the diametral point load mean value is computed as 1.89. It is expected that, due to the bedding nature of sedimentary deposits, in rock testing on vertical cores, the diametral point load test will be vulnerable to the weakening bedding effects and will yield lower strength values compared to axial point load tests, i.e., strength anisotropy. However, both point load test modes are correlated with σ_{ci} values on vertical cored specimens. In the context of concrete-rock joint interaction in a foundation rock socket, it is the lower rock mass strength in horizontal direction that is mobilized to resist radial dilation and circumferential tensile strains induced by foundation socket loading. When using empirical models for rock socket design, consideration should be given to this strength anisotropy effect.

Pells [4] has reported a range from 14 to 35 with a best-fit value of 24 for diametral point load value conversion to UCS on saturated samples, compared with a conversion factor of 37 in this study, (see Fig. 3 upper plot).

The modulus ratio (MR) values were found to be with a mean of 107, std. dev. of 5, with a range of 47–209. These values are generally below the results shown in the MR plot in [4]. They are well below the MR range quoted in [8] for sandstone,

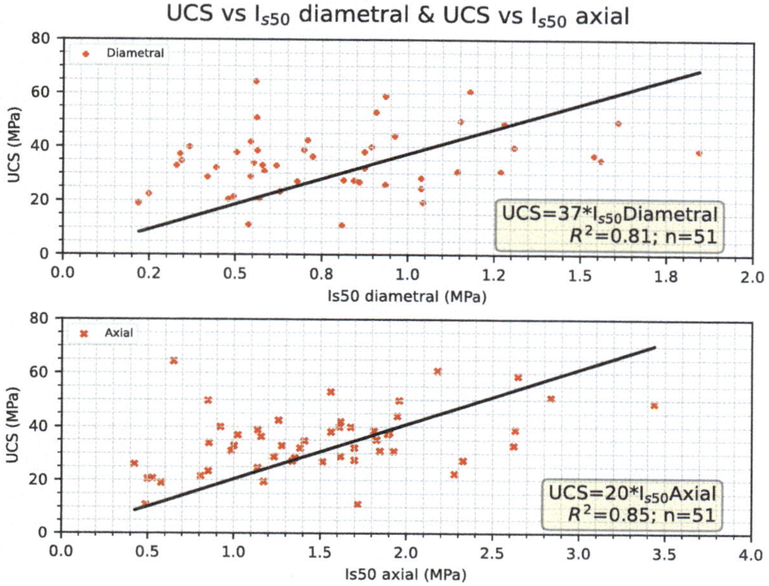

Fig. 2 σ_{ci} (UCS) versus I_{s50} transformation model

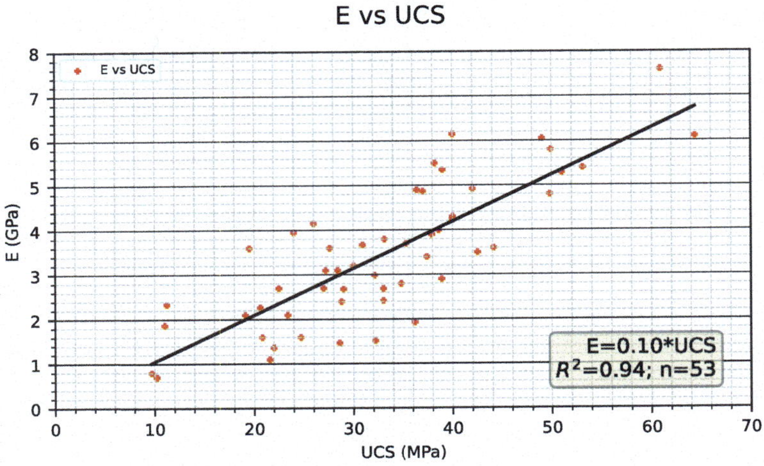

Fig. 3 E_{i} versus σ_{ci} transformation model

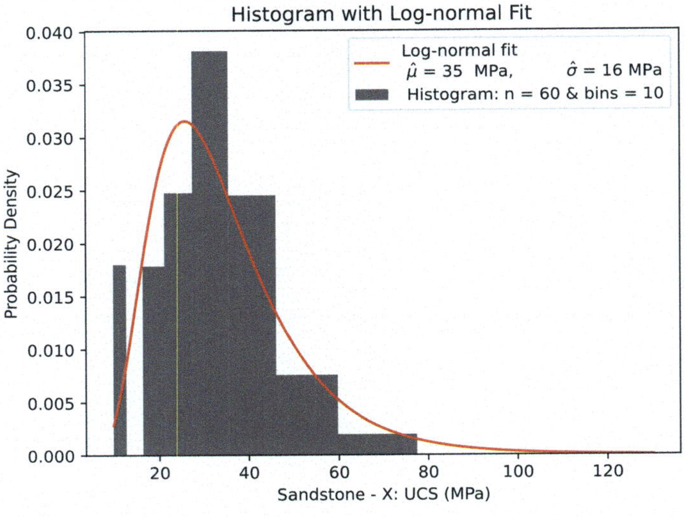

Fig. 4 Frequency density plot of σ_{ci}, (UCS) data with the fitted lognormal distribution

i.e., 200–350. The low E_{i} values (Table 1) and the corresponding MR values will be further discussed in the paper.

4.2 *Inferential Statistics of the Sandstone Database*

Fitting a probability distribution: In view of the non-negative nature of the data, σ_{ci}, values of the sandstone database, was fitted with a lognormal distribution using the method of moment estimators [9] (Fig. 4).

Goodness-of-fit tests: At the 5% level of significance, the null hypothesis, i.e., the data fits the lognormal distribution cannot be rejected with the Kolmogorov–Smirnov (K-S) test. This test is considered good for identifying discrepancies across the entire distribution but can be less sensitive to differences in the tails. On the other hand, the Anderson–Darling (A-D) test gives more weight to the tails of the distribution. This test, at the same significance level, yields a statistic that exceeds the critical value and hence falls into the rejection region, i.e., the data does not support a fit to the lognormal distribution. However, at a level of significance of 1%, it fails to reject the null hypothesis.

Statistical analysis needs to be looked at in the context of geological information. Table 1 indicates the subject σ_{ci} data are classified in the upper end of the strength classification. A-D results are more weighted to the tails of the probability distribution. Then the tail of the pdf distribution that impacts the subject test data is on the right-hand side, i.e., the region of higher strengths. The percentage of intact strengths higher than 48 MPa is less than 20%. Although it can have an impact on limiting socket interface roughness, it is on the better side of the pdf distribution, indicative of higher intact strengths.

5 Rock Mass Characterization

5.1 *Rock Mass Modulus,* E_m

There are several approaches in the literature to get preliminary estimates of E_m such as based on σ_{ci} [10], and GSI (Geotechnical Strength Index is a rating based on the subjective assessment of lithology, structure, and surface conditions of the rock mass discontinuities) based [8], among others. Following Johnston et al. [11] who report that the rock mass moduli back-calculated from pile load test data were found to be compatible with rock pressuremeter data, such project pressuremeter data obtained in this investigation are reviewed.

5.2 *Rock Pressuremeter Testing*

Rock pressuremeter testing was carried out targeting different lithologies. Twenty pressuremeter tests were undertaken in six boreholes targeting sandstone/sandstone dominant beds and only the sandstone results are discussed. Fifteen test locations had both σ_{ci} and E_i test data very nearly at the same elevations. There were few natural infrequent defects in the rock mass and these defects were generally tight and largely lack of infillings. At the test elevations, the mean RQD was 87% with a standard deviation of 14%. GSI values at the test locations ranged from 50 to 70, mostly between 65 and 70. All testing was performed in NMLC size cored boreholes; the common problems in pressuremeter testing due to fractured rock and sheath rupture were experienced at a few locations.

All tests were conducted under stress-controlled conditions up to a maximum pressure of 10 MPa or terminated earlier on reaching the pressuremeter limits. Typically, three load/unload loops were performed at different test pressures. The loading–unloading stress magnitude was not greater than about 1.5 MPa. The response of the rock to applied radial pressure (cavity pressure) at the cavity wall was measured as average displacement (using a point strain measuring pressuremeter) and converted to radial strain. Test results were reported in the cavity pressure (p)–cavity strain (circumferential strain, ε_c) space. Due to potential disturbance during probe insertion, the only parameter that can be estimated with reliability from the initial elastic portion of the curve is the shear modulus. A Poisson ratio of 0.25 was used to evaluate the elastic modulus, E_{pi}. The p–ε_c plots were of the elastic and strain hardening type with no discernible evidence of tensile failure [12]. Unload-reload loops indicated very high elastic stiffness and were reported as exceeding 10 GPa. Hence, progression of fracture development could not be identified from the plots from successive unload-reload loops. It is argued that rock socket loading in the rock mass environment where the surrounding rock has been disturbed will likely be more related to the initial tangent stiffness from pressuremeter testing.

5.3 Comparison of Initial Pressuremeter Modulus E_{pi} with Empirical Predictions

In Fig. 5, E_{pi} is the pressuremeter initial tangent modulus, E_i is the intact rock modulus, and the cavity strain, $\varepsilon_c = (r - r_0)/r_0$. r_0 is the initial cavity radius, and r is the cavity radius under cavity pressure p. The E_{pi} values associated with the initial tangent points varied from 800 to 9000 MPa, with a mean of 4325 MPa and COV of 0.53.

This appears to be an anomalous result with the mass modulus exceeding the intact modulus. Pells [4] reports E_m (mass modulus) for Class I/II Sydney sandstone between 900 and 2500 MPa. As discussed in Sect. 4.1, the reported σ_{ci} values would suggest that the investigated sandstone to be stronger than in [4; Sydney sandstone] but the E_i values to be significantly smaller, by about a factor of about 2. Further, as stated earlier, the unload-reload moduli akin to uncracked moduli, measured in the pressuremeter tests exceeded 10 GPa. Notwithstanding the stress relief-induced potential micro-cracking on retrieved intact core samples, the E_i values should bear

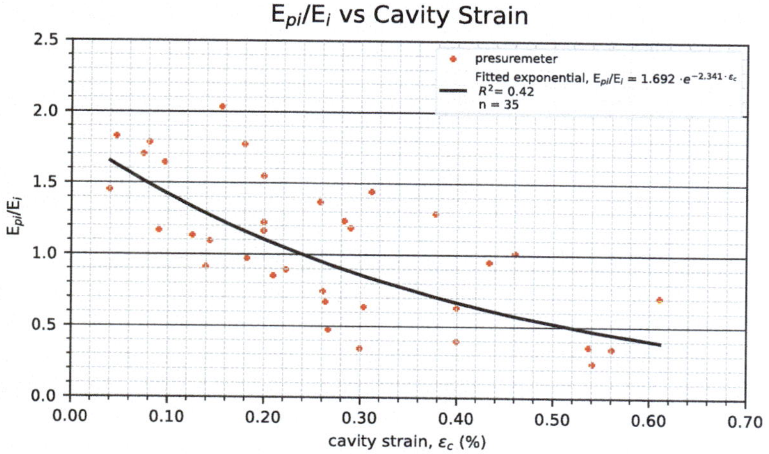

Fig. 5 Normalized rock pressuremeter initial stiffness with cavity strain

close resemblance to unload-reload moduli in pressuremeter tests. This could be one probable explanation for the anomalous high E_{pi}/E_i ratios (due to possible normal faulting induced stress-relief impacts on E_i). There is limited evidence cited in [8] about some Italian research where higher than intact moduli have been reported for mass modulus for non-foliated rocks. Notwithstanding the above, strict caution should be exercised to ensure mass modulus values do not exceed intact values and recommendations for E_m/E_i given in [8] should serve at least for preliminary design. For the GSI range reported in this study, i.e., 50–70, the computed values for the ratio E_m/E_i are approximately between 30% and 70%, with increasing GSI, according to [8].

6 Direct Shear Testing of Concrete-Rock Interface Joints

6.1 *Influence of Socket Roughness*

One of the significant factors that influence shaft resistance and well recognized is socket roughness. In excavation of rock sockets in weak rocks, the pile rock interface of drilled sockets generally consists of irregular patterns and heights of asperities e.g., [6]. Mobilization of side-shear resistance at the pile (concrete) rock interface enables load transfer when sockets are under axial loading. The load transfer mechanism is brought about by processes such as sliding on irregular surfaces, concrete asperities (pushed downwards by the axial loading) overriding rock asperities. This, in turn, results in the wearing and degradation of asperities (see Fig. 6). In weak rock masses, concrete will be stronger (having a higher σ_{ci}) than the surrounding rock mass.

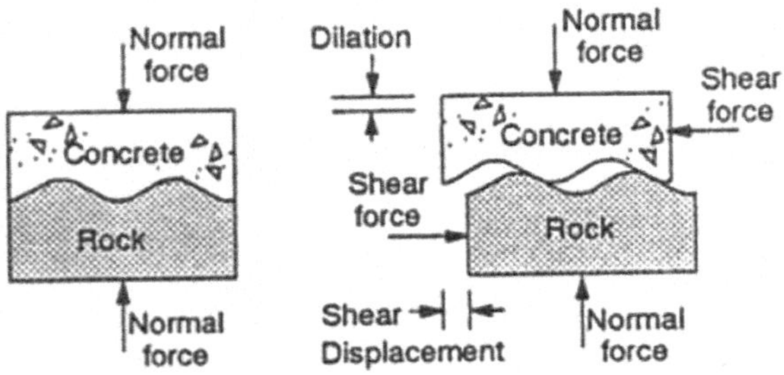

Fig. 6 Concrete-rock joint interface sliding (schematic)

For simplicity, assuming the concrete pile to be rigid, the lateral pushing of the rock asperities by the axially loaded rigid shaft will be progressively resisted by surrounding rock mass depending on the lateral stiffness of the surrounding rock. This boundary constraint, results in producing constrained dilatancy. Therefore, the mobilized boundary condition results in an increase in the normal stress, $\Delta\sigma_n$, at the concrete-rock, socket interface.

This constrained dilatancy effect can be approximately modeled as an expanding cylindrical cavity in an elastic infinite medium [13], as commonly adopted in rock-socket literature (see Eqs. 1 and 2). In the cavity expansion problem, a radial element into the socket wall experiences a principal stress at the socket wall. Further, the elasticity-based cavity expansion problem, without the ability to impose strength cut-offs, cannot simulate radial cracking brought about by the tensile circumferential strains due to the significantly lower tensile strength of the surrounding rock mass, compared to its compressive strength.

In the corresponding interface socket shear problem, a radial element is subject to a normal stress/shear stress condition at the socket wall. Despite the differences, insightful information can be expected to be obtained from the cavity expansion problem away from the socket wall. To simulate this boundary condition, the ratio of the increase in normal interface stress and interface dilation is maintained constant during the test, and the ratio is known as constant normal stiffness, K_N, and was defined in [11] as follows:

$$\Delta\sigma_n = E_m/(1 + \nu_m) * \Delta r_s/r_s = 2G_m * \Delta r_s/r_s \quad (1)$$

$$K_N = \Delta\sigma_n/\Delta r_s = 2G_m/r_s \quad (2)$$

where $\Delta\sigma_n$ is the change in normal stress brought about by the constrained dilation of the concrete-rock socket interface; and Δr_s, E_m, G_m, and ν_m are the socket interface dilation, deformation modulus, shear modulus, and Poisson's ratio, respectively, of the surrounding rock mass. In a conventional direct shear test, the normal load is maintained constant during shear (hence called the constant normal load test (CNL)). To simulate the mobilized lateral resistance of the surrounding rock mass on the expanding cavity of the rock socket under load in the laboratory, the normal load is increased in such a way to maintain $\Delta\sigma_n/\Delta\delta_v$ constant ($\Delta r_s = \Delta\delta_v$, the normal displacement in the direct shear tests). This type of test is called the constant normal stiffness test (CNS).

6.2 CNS and CNL Rock Direct Shear Testing

Sandstone and sandstone dominant rock cores were subject to concrete-rock joint testing under CNS and CNL test boundary conditions at Monash University in Melbourne. As input parameters for this testing, effective roughness heights, Δr (to develop the concrete-rock joint interface), and rock mass stiffness estimates (to determine the range of constant normal stiffness, K_N) are required.

As we are dealing with low to very low strength rocks, i.e., σ_{ci}, predominantly less than 50 MPa, rock socket construction is mainly undertaken with rock augers [14]. As a result, asperities are formed on the socket walls. Field-based estimate bounds of the roughness heights are available in the literature, e.g., [6]. In the Monash socket roughness model, the interface roughness profiles are modeled as a series of interconnected chords of equal length using fractal concepts (Table 2).

The following observations based on the computed parameters in Table 3 can be made:

(i) CNL results show a "brittle" response, i.e., significantly small shear displacements at peak shear stress in contrast to the ductile behavior exhibited by the CNS tests (see also Fig. 7). Pells [15] has also shown that concrete/rock interface shearing with low confinement (akin to CNL conditions) produces peak

Table 2 Summary of CNS/CNL concrete-rock joint test details[a]

Test/mean asperity height, Δr (mm)/mean asp. angle, α (°)/std. dev. of α (°)$^+$; [σ_{ci}] (MPa)	Constant normal stiffness, K_N (kPa/mm)	Peak shear stress, τ_p (kPa)	Shear displacement, δ_s at τ_p (mm)
CNS_C1: 2.2/8/10.1; [36]	800	1358	14.4
CNS_C2: 2.2/8/10.1; [36]	1000	1540	9.8
CNL_C1: 2.2/8/10.1; [11]	0	634	0.8
CNL_C2: 2.2/8/10.1; [11]	0	597	2.2
CNS_B1: 1.4/5/6.3; [31]	800	1124	14.0
CNS_B2: 1.4/5/6.3; [31]	1000	1440	13.8

Notes [a] Carried out on sandstone or sandstone dominant intact samples
Initial normal stress, $\sigma_{no} = 600$ kPa for all the tests. $^+$ Constant chord length = 16 mm

Table 3 Analysis of CNS/CNL concrete-rock joint test results[a]

Test	σ_{ci} (MPa)	Peak shear stress, τ_p (kPa)	τ_p/σ_{ci}	K_N (kPa/mm)	δ_s at τ_p (mm)	σ_N at τ_p (kPa)	τ_p/σ_N	ϕ'^{b} (°)
CNS_C1	36	1358	0.038	800	14.4	1687	0.80	38.8
CNS_C2	36	1540	0.043	1000	9.8	2012	0.77	37.4
CNL_C1	11	634	0.058	0	0.8	600	1.06	46.6
CNL_C2	11	597	0.054	0	2.2	600	1.00	44.9
CNS_B1	31	1124	0.036	800	14.0	1712	0.66	33.3
CNS_B2	31	1440	0.046	1000	13.8	2183	0.66	33.4

[a] The C-series based on $\Delta r = 2.2$ mm, and the B-series based on $\Delta r = 1.4$ mm; initial normal stress, $\sigma_{no} = 600$ kPa for all the tests
[b] Secant friction angle

shear stresses at very low shear displacements. Gu et al. [16] have shown similar trends based on CNS/CNL tests carried out on triangular asperities. This will translate to an early mobilization of sideslip in the field in the presence of low confining stiffness of the surrounding rock mass. Overall, the stiffness of the surround rock mass has a significant influence on both settlements of rock sockets as well as producing a ductile response.

(ii) CNL results are associated with higher ratios of τ_p/σ_N, i.e., higher secant friction angles, in comparison with the values for the CNS test. This is akin to an over-consolidated response in soils under low normal stresses.

6.3 Comparison of CNS/CNL Results with σ_{ci}-Based Empirical Correlations for Peak Shear Stress, τ_p

The following observations can be made on the computed parameters in Table 4 which are based on the CNS/CNL direct shear testing:

(i) The closest prediction to the observed peak shear stresses is given by the WJD [17] correlation. It must be noted that the coefficient "*a*" for this method, unlike the other methods investigated, is a function of UCS, but certainly it does not require an extra input in the application of the correlation.

(ii) The highest overprediction of all the observed peak shear stresses in this study is given by the RA [10] correlation.

(iii) All the reported σ_{ci}-based correlations in Table 4 overpredicted the peak shear stresses for the CNL specimens. This overprediction could be likely because the field-calibrated σ_{ci}-based correlations do reflect in-situ rock confinement effect and hence may not directly simulate the CNL low normal stress boundary condition.

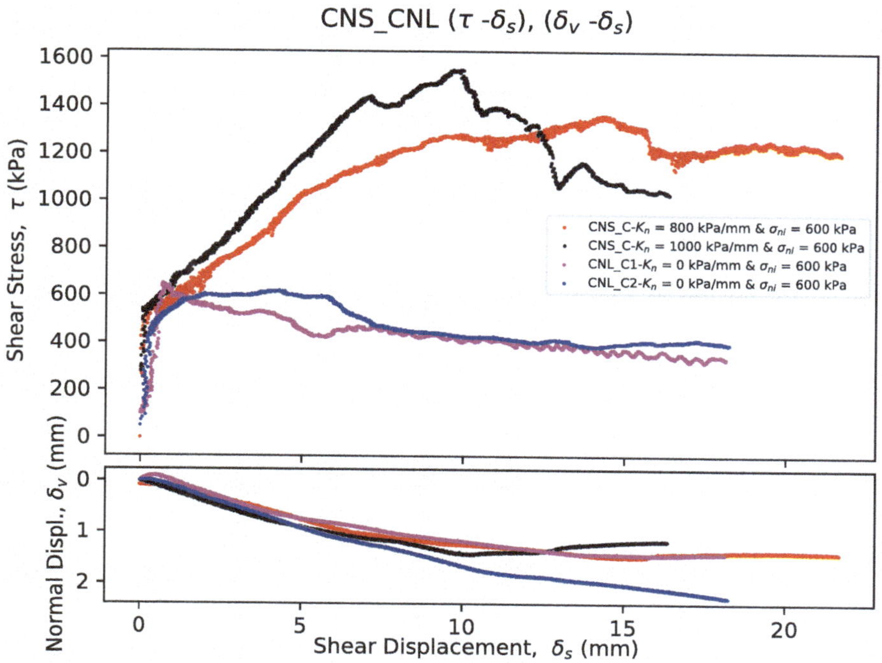

Fig. 7 CNS/CNL concrete-rock joint direct shear test results

Table 4 Analysis of CNS/CNL concrete-rock joint test results

Test	σ_{ci} (MPa)	Peak shear stress, τ_p (kPa)	Ultimate shaft friction (kPa) $f_s/p_a = a * [\sigma_{ci}/p_a]^b$; $b = 0.5$			
			WJD [17] $a = f(\sigma_{ci})$ $b = 1$	CK [18, 19] $a = 1$	RA [10] $a = 1.42$	ZE [20, 21] $a = 0.95$
CNS_C1	36	1358	$a = 0.040$ 1440	1897	2694	1800
CNS_C2	36	1540				
CNL_C1	11	634	$a = 0.093$ 1023	1049	1489	995
CNL_C2	11	597				
CNS_B1	31	1124	$a = 0.050$ 1550	1761	2500	1670
CNS_B2	31	1440				

6.4 Comparison of CNS Results with Reported Test Pile Information

Reference [1] has reported one of the Osterberg Cell test pile results. For significant projects, such tests are important as they can simulate the true field conditions as close as possible. Figure 3 of [1] shows a peak side-shear resistance of approximately 1750 kPa mobilized at an average shear displacement of approximately 13 mm.

The approximate average peak shear stress for CNS_C series was 1450 kPa and for CNS_B series 1280 kPa. The corresponding shear displacements were 12.1 mm and 13.9 mm, respectively. Hence, the rock direct shear CNS tests have provided safe lower bound results confirmed by static test pile load test results.

7 Significance of Pile Group Effects

It was mentioned earlier that to carry out a rational design of an individual socket, and it is essential to have a tolerable settlement criterion for the pile group. Design predictions for the First Bridge [2], including rock/soil/foundation/superstructure interaction using elasto-plastic boundary element methods, had predicted a maximum group settlement of 40–50 mm. Corresponding maximum single pile settlements of 5–12 mm were also predicted. A back-of-the-envelope calculation was undertaken using the settlement ratio method [24] to predict the single pile settlements in the group based on the group settlements predicted in [2]. This resulted in single pile settlements of 6 mm and 7 mm, respectively.

The magnitudes of the working settlements computed above indicate that the rock socketed foundations predominantly carry the axial load in side-shear and more importantly single rock socket performance is likely to be dictated by settlements. This points to the importance of advancing settlement-based rock socket foundation design methods. Recent work in this area, for example, is discussed in [23, 17].

8 Summary of Main Findings

This paper has clearly demonstrated the value and the insight provided by the pre-tender geotechnical investigations for significant projects to provide input on ground characterization. This study has been limited to the sandstone lithology. Opportunity has been taken to review the findings in the light of recent research and understanding of the subject. No attempt has been made to discuss design-specific code issues [22]. The importance of sound construction practices is particularly significant for aerially constricted structures such as at bridge abutments, i.e., transportation projects. At these abutment locations, the foundation elements need to be constrained within the headstock and afford little room for redundancy. These issues take critical importance in rock socket foundations.

The main findings can be summarized as follows:

In-situ stress field: The influence it has on rock mass characterization, and on cavity expansion type of loading in deep rock sockets by enhancing the resistance against radial cracking. However, how the substance properties of vertically cored specimens are influenced due to possible stress relief effects in natural stress fields with different major principal stress directions need further attention.

Rock substance characterization: The most accessible type of information and most empirical rock engineering design methods are based on rock substance properties. This role should not be overlooked as mass properties are related to the substance properties. Rock strength anisotropy identified should be given careful consideration.

Benefits of CNS/CNL rock direct shear testing: By simulating the confining effect of the surrounding rock mass, it has shown the significant increase in shear resistance in the CNS test compared to the CNL test. This also raises the importance of characterizing the discontinuity nature of the rock mass.

Rock pressuremeter testing: Some difficulties for interpretation have been identified, and probable reasons advanced for further confirmation. With the advances in electronics and measurement accuracies [12], rock pressuremeter testing should play a greater role in understanding the stiffness of the rock masses in response to the paradigm that more emphasis needs to be made on settlements for rock socketed foundation design. Finally, the findings need to be confirmed by test piles.

Acknowledgements The authors would like to acknowledge the contribution given by our colleagues, Jim Martin (ex-Principal Geologist), David Marks, Peter Reynolds, and Brad Woodgate during the pre-tender project phase. Dr. Jay Kodikara (Monash University) for undertaking the rock direct shear testing at Monash, in consultation with TMR. Dr. Jay Ameratunga (T+T) for his valuable review of the paper. The first author was the Department's lead geotechnical investigator of the reported study. The views expressed in this paper are the views of the authors and do not necessarily reflect the views of the Department of Transport and Main Roads, Queensland.

References

1. Day R, Johnston I, Yang D (2014) Design of foundations to second gateway bridge—Brisbane. In: Proceedings of the 7th AustRoads bridge conference, Adelaide
2. Williams DJ (1989) Geotechnical input to major bridge project. ASCE J Geotech Eng 115(3):322–339
3. Willmott W, Stevens N (2012) Rocks and landscapes of Brisbane and Ipswich, 2nd edn. Geological Society of Australia, QLD Division
4. Pells PJN (2004) Substance and mass properties for the design of engineering structures in the Hawkesbury Sandstone. Aust Geomech 39(3):1–21
5. Hencher S (2015) Practical rock mechanics. CRC Press
6. Siedel JP, Collingwood B (2001) A new socket roughness factor for prediction of rock socket shaft resistance. Can Geotech J 38:138–153
7. Ching J, Li K-H, Phoon K-K, Weng M-C (2018) Generic transformation models for some intact rock properties. Can Geotech J 55(12):1702–1741
8. Hoek E, Diederichs MS (2006) Empirical estimation of rock mass modulus. Int J Rock Mech Min Sci 43:203–215
9. Fenton GA, Griffiths DV (2008) Risk assessment in geotechnical engineering. Wiley
10. Rowe RK, Armitage HH (1987) A design method for drilled piers in soft rock. Can Geotech J 24(1):126–142

11. Johnston IW, Lam TSK, Williams AF (1987) Constant normal stiffness direct shear test for socketed pile design in weak rock. Geotechnique 37(1):83–89
12. Hughes J, Whittle W (2022) High resolution pressuremeters and geotechnical engineering: the measurement of small things. CRC Press
13. Yu HS (2000) Cavity expansion methods in geomechanics. Springer
14. Turner J (2006) NCHRP synthesis 360: rock-socketed shafts for highway structure foundations. TRB, Washington, DC
15. Pells (1980) An experimental investigation into side shear for socketed piles in sandstone. In: Proceedings of the international conference on structural foundations on rock, Sydney
16. Gu XF, Seidel JP, Haberfield CM (2003) Direct shear test of sandstone-concrete joints. Int J Geomech 3(1):21–33
17. Johnston IW (2022) Evaluation of design methods for side-resistance-only rock-socketed piles. Can Geotech J 59:601–613
18. Carter JP, Kulhawy FH (1988) Analysis and design of drilled shaft foundations socketed into rock. Report EL-5918. Electric Power Research Institute, Palo Alto, CA
19. Kulhawy FH, Prakoso WA, Akbas SO (2005) Evaluation of capacity of rock foundation sockets. In: Proceedings of the 40th U.S. rock mechanics symposium, Anchorage, Alaska
20. Zhang I, Einstein H (1998) End bearing capacity of drilled shafts in rock. J Geotech Eng ASCE 124(7):574–584
21. Small JC (2016) Geomechanics in soil, rock, and environmental engineering. CRC Press
22. Look B, Lacey D (2013) Characteristic values in rock socket design. In: Proceedings of the 18th international conference on soil mechanics and geotechnical engineering, Paris
23. Haberfield CM, Lochaden ALE (2019) Analysis and design of axially loaded piles in rock. J Rock Mech and Geotech Eng 11:535–548
24. Poulos HG (2017) Tall building foundation design. CRC Press

Recent Advances in the Use of Artificial Intelligence for Rigid Airfield Pavement Analysis and Design

Halil Ceylan, Orhan Kaya, and Sunghwan Kim

Abstract Historically, analysis and design methodologies for rigid airfield pavement systems have primarily relied on empirical equations derived mostly based the field performance. However, the FAA has implemented mechanistic-based approaches, including three-dimensional finite-element (3D-FE) procedures, for the design and analysis of rigid airfield pavements. This decision was made in response to the emergence of new wide-body aircraft and other design complexities, such as heavier aircraft and intricate gear configurations. On the other hand, the time-consuming nature of 3D-FE computation for analyzing multiple slabs subjected to aircraft and environmental loads has rendered routine design and analysis impractical. To address this issue, AI-based alternatives offer significant potential for producing accurate and rapid rigid pavement-response estimations compared to traditional FE-based design programs. Using AI-based modeling is a convenient option instead of conducting lengthy 3D-FE computations. In this paper, we introduce a recent FAA-sponsored research study conducted at Iowa State University (ISU) that utilizes AI-based alternatives to performing full 3D-FE computation for analyzing and designing rigid airfield pavement systems. The paper contains two case studies: (1) creating strong AI models to predict critical rigid airfield pavement responses and analyze top-down cracking behavior and (2) developing fast AI-based models to predict pavement foundation response and moduli for designing new and rehabilitated rigid airfield pavement structures. The capability of AI-based surrogate-response models for analyzing and designing rigid airfield pavement systems is successfully demonstrated and discussed in this paper.

Keywords Artificial intelligence · ANN · FAARFIELD · Rigid airfield pavement design

---

H. Ceylan (✉) · S. Kim
Department of Civil, Construction and Environmental Engineering, Iowa State University, Ames, IA 50011, USA
e-mail: hceylan@iastate.edu

O. Kaya
Department of Civil Engineering, Adana Alparslan Türkeş Science and Technology University, Adana, Turkey

B. Indraratna and C. Rujikiatkamjorn (eds.), *Recent Advances and Innovative Developments in Transportation Geotechnics*,
https://doi.org/10.1007/978-981-97-8245-1_6

1 Introduction

Airfield pavements have been designed and constructed to resist repeated loads placed on them by aircraft as well as other negative impacts such as unfavorable weather (such as extremely hot or cold weather). Moreover, they are expected to serve a fleet of airplanes with varying weights and gear layouts. As a result, airfield pavements must be built to endure years of continuous traffic loads from all types of aircraft, not just the heaviest. Rigid airfield pavements, composed of concrete slabs at the top and the pavement foundation beneath the slabs, have been one of the most used airfield pavement types. Historically, while design and analysis methodologies for rigid airfield pavement systems have primarily relied on empirical research as well as field performance, the emergence of new wide-body aircraft along with other design challenges, such as increased aircraft weight and more complicated gear configurations, has prompted the Federal Aviation Administration (FAA) to use mechanistic-based methodologies with three-dimensional finite-element (3D-FE) procedures as part of the rigid airfield pavement analysis and design. The recently implemented mechanistic-empirical design methods, integrated into the FAA's Rigid and Flexible Iterative Elastic Layer Design (FAARFIELD) program, are robust and flexible enough to handle complex gear configurations that may arise in future without the need to alter the fundamental procedures [1]. FAARFIELD utilizes the NIKE3D_FAA 3D-FE computer program to calculate the maximum horizontal stress at the bottom edge of a Portland Cement Concrete (PCC) slab, which is the critical pavement response considered in rigid airfield pavement design. This prediction helps determine the structural life of the pavement. The NIKE3D software was originally developed by the Lawrence Livermore National Laboratory (LLNL), a part of the US Department of Energy, and was later adapted for use in FAARFIELD [1].

The current FAA rigid pavement design model only considers a single slab loaded at one edge by a single aircraft gear and does not effectively address top-down cracking. As a result, the FAARFIELD rigid airfield pavement design procedure does not adequately account for this failure mode. Additionally, FAARFIELD does not currently factor in the failure of subgrade and subbase layers for rigid airfield pavements and only calculates slab thickness based on bottom-up cracking as the mode of failure for a specific aircraft traffic mix [2].

The FAA has created finite-element analysis-FAA (FEAFAA), a separate application that uses NIKE3D to analyze rigid airport pavements with multiple slabs through 3D-FE analysis. It accurately calculates pavement responses, including stresses, deflections, and strains, under specific airplane landing gear loads. Although FEAFAA is a useful tool for research and analysis, it lacks the transfer function elements of FAARFIELD and cannot be used as a complete design tool.

As previously mentioned, the current design methodology in FAARFIELD only considers bottom-up cracking as a failure criterion and does not sufficiently address top-down cracking. To address this issue, a research version (FAARFIELD 1.4-TDC) of the FAARFIELD design software has been created. This new version replaces the single-slab 3D-FE response model with a four-slab 3D-FE model that considers variable joint spacing and initial temperature curling. However, routine design using this updated software was found to be impractical due to the lengthy and unpredictable processing times of the four-slab model and curled slabs. The FAA is looking for more efficient options for conducting 3D-FE stress calculations without the need for client software. Additionally, FAARFIELD currently only takes into account the maximum horizontal stress at the bottom edge of the concrete slab when determining bottom-up cracking failure of the PCC layer, without considering the critical responses for failure of subgrade and subbase layers in rigid pavement designs. It is evident that integrating both top-down cracking and pavement foundation responses into airfield pavement design processes would be highly beneficial.

In response to this problem, artificial intelligence (AI) solutions like artificial neural networks (ANNs) show great promise in providing precise and quick estimates of rigid pavement response, surpassing those generated by conventional FE-based design software. ANNs have been chosen as the preferred method [3, 4] for creating an alternate model using AI approaches, offering a rapid and more reliable alternative to the complex and time-consuming FE-based structural modeling. This approach has been successfully applied in the analysis and design of airport and highway concrete pavements [2–4]. The adoption of surrogate-response modeling has offered rapid and precise solutions for different input parameters, which have been applied in the creation of the Mechanistic-Empirical Pavement Design Guide (MEPDG), currently referred to as AASHTOWare Pavement ME [5]. Artificial neural networks (ANNs) have been selected for their adaptability and demonstrated effectiveness in stand-in pavement-response modeling.

In this paper, we introduce a recent FAA-sponsored research study conducted at Iowa State University (ISU) [6–10] that utilizes AI-based alternatives to perform a full 3D-FE computation for analyzing and designing rigid airfield pavement systems. The paper includes two case studies: developing robust AI models that predict critical rigid airfield pavement responses and analyze top-down cracking behavior [6–8] and developing rapid AI-based pavement foundation moduli and response prediction models for designing new as well as rehabilitated rigid airfield pavement structures [9, 10]. The capability of AI-based surrogate-response models for analyzing and designing rigid airfield pavement systems is successfully demonstrated and discussed in this paper.

2 Development of Robust AI-Based Models to Predict Critical Rigid Airfield Pavement Responses and Analyze Top-Down Cracking Behavior

The goal of this study was to create a substitute AI-based response model or process (appropriate for use in FAARFIELD 1.4-TDC) that can provide a similar estimation of the top-down bending stress calculated by NIKE3D for both combined mechanical and temperature loading and mechanical-only loading in rigid airport pavements. This model will enable faster 3D-FE calculations of design stresses, making FAARFIELD 1.4-TDC more practical for routine design applications [6–8]. The developed models were used in a separate tool known as ANNFAA, which is designed to be integrated into the upcoming FAA's thickness design software, named FAARFIELD 2.0. This tool also provides a reliable approximation of the top-down bending stress calculated by NIKE3D_FAA for both combined mechanical and temperature loading and mechanical-only loading of rigid airport pavements [6]. The new ANNFAA tool, developed as part of this study, can be effectively utilized in conjunction with FAARFIELD 2.0 to greatly decrease the amount of time needed to complete a design analysis for FAA rigid pavements. The detailed procedure of the model development stages and the details of the ANNFAA could be found in the publications produced by the ISU team [6–8]. Rezaei-Tarahomi et al. [6] illustrated the potential benefits of incorporating the findings of this study into the FAARFIELD 2.0 rigid airfield pavement design framework in a flowchart, and a simplified version of this flowchart is presented in Fig. 1. As can be seen in Fig. 1, instead of relying on 3D-FE analysis, the use of ANNFAA could effectively predict critical stresses, specifically the maximum horizontal stresses at the slab surfaces. These predicted stresses could then be utilized to calculate other necessary parameters.

The methodology for this study includes the following steps: (1) generating a synthetic database with FEAFAA 2.0 using a nine-slab model; (2) using ANN methodology to create a predictive model for top-down bending stress by mapping nonlinear input–output relationships in the database; (3) confirming the accuracy and reliability of the predictive model; (4) assessing and validating the model; (5) creating a new ANN tool for analyzing rigid airfield pavements with multi-slab (nine-slab) models to accommodate all aircraft types in FEAFAA 2.0.

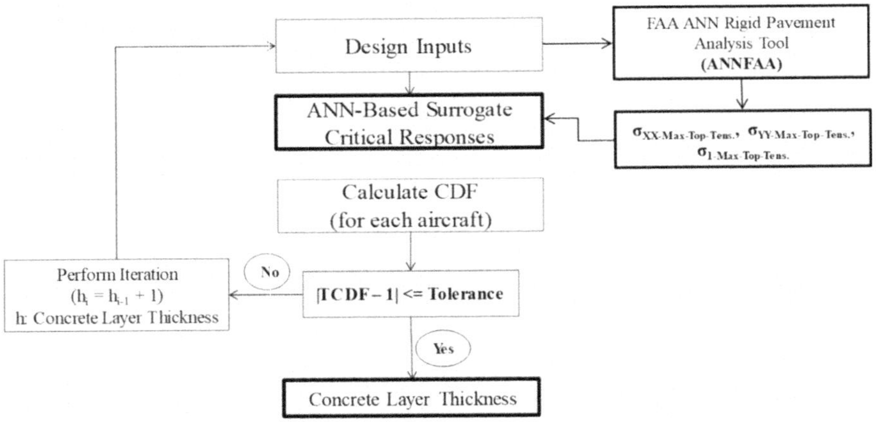

Fig. 1 Flowchart displaying the contribution of this study in FAA rigid airfield pavement design (adapted from [6])

The ISU project team created a batch run automation tool with the capabilities of 3D-FE computation along with post-processing the outputs. This tool allowed researchers to create a database composing of input–output records used in NIKE3D-FAA FE computation for each aircraft type and loading scenario, obtaining principal and tensile stresses in both x and y directions. Additionally, a MATLAB tool was developed to streamline the creation of ANN models, improving data import, model training, and export capabilities. This tool facilitated the development of 6000 ANN models for two loading conditions for each aircraft type, using various architectures and training algorithms. ANN models were created for all 156 aircraft types in FEAFAA 2.0.

ANNFAA enables quick determination of essential pavement responses. It takes just 18.8 s to extract inputs, bias, and weight matrices from text files, and an additional 6.3 s to calculate critical responses for 1 million cases. This makes ANNFAA a valuable tool for efficiently conducting standard rigid airfield pavement design tasks [6].

3 Development of Rapid AI-Based Pavement Foundation Response and Moduli Prediction Models for Designing Both New and Rehabilitated Rigid Airfield Pavement Structures

As previously mentioned, the present FAARFIELD rigid airfield pavement design procedure only considers the maximum horizontal stress at the bottom edge of the concrete slab to determine bottom-up cracking failure of a PCC layer. However, it does not consider the critical responses for failure of subgrade and subbase layers, even though it is clear that pavement foundation-related responses should be included in the design process. ISU team developed rapid 3D-FE-based pavement foundation response as well as moduli prediction models for rigid airfield pavement structures [9, 10]. Specifically, the models were created for three case studies: predicting rigid pavement foundation responses for various wide-body aircraft loading conditions, predicting responses for heavy/falling-weight deflectometer (H/FWD) loading conditions, and creating a single rigid pavement moduli prediction model. These prediction models were based on ANNs and provided a close estimate of the responses calculated by NIKE3D employed in FAARFIELD. The details of the rapid 3D-FE-based pavement foundation response as well as moduli prediction models for rigid airfield pavement structures, developed by the ISU team, can be found in Rezaei-Tarahomi [9] and Rezaei-Tarahomi et al. [10]. The overall approach used by ISU team for developing rapid 3D-FE-based pavement foundation response as well as moduli prediction models for rigid airfield pavement structures is illustrated in Fig. 2. First, to create the ANN models, a database was created with randomly generated FEAFAA input

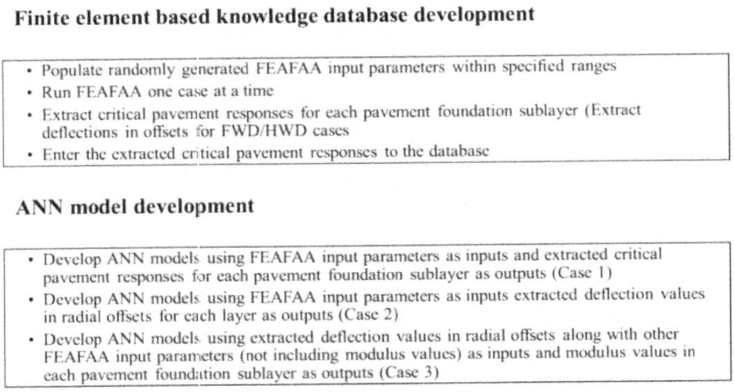

Fig. 2 Overall description of ANN model development approach (adapted from [9, 10])

parameters within set limits and corresponding pavement responses computed by FEAFAA for each layer. The relationships between the input and output parameters were then used to develop the ANN models based on the data in the database.

In Case 1, individual ANN models were created to estimate critical pavement responses in each sublayer of the pavement foundation (Fig. 3a). Separate ANN models were also developed for two different wide-body aircraft loading scenarios: a Boeing B777-300ER and an Airbus A380-800. All of the ANN models were developed successfully predicted critical pavement responses, with each model achieving a correlation coefficient (R^2) value of 0.999 or higher.

In Case 2, FWD/HWD testing was simulated in FEAFAA, and ANN models were created with FEAFAA input parameters and deflection values in radial offsets as outputs (Fig. 3b). The radial offsets selected for analysis were at distances of 0 mm, 305 mm (12 in), 610 mm (24 in), 914 mm (36 in), 1219 mm (48 in), and 1524 mm

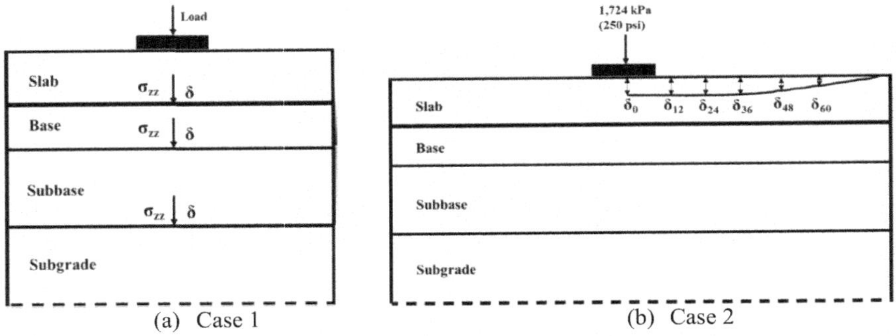

Fig. 3 Pavement responses predicted using developed ANN-based models for **a** Case 1 and **b** Case 2 (adapted from [9, 10])

(60 in) from the load center to mimic real HWD loadings on the pavement system. The ANN models accurately predicted the FEAFAA results of overall deflections at these six points, with an R^2 value of 0.999 or higher.

In Case 3, the goal was to determine the elastic modulus values for the pavement layers through backcalculation. An ANN model was created to calculate the elastic modulus of all layers using deflection values at radial offsets and other input parameters (excluding modulus values) as inputs and elastic modulus values of each pavement foundation sublayer as outputs. The advantage of using this ANN backcalculation model in routine HWD/FWD evaluations is the ability to quickly process and analyze data, possibly using them in the field to obtain real-time results.

4 Conclusions

In this paper, we introduced a recent FAA-sponsored research study conducted at Iowa State University (ISU) that utilizes AI-based alternatives to perform full 3D-FE computation for analyzing and designing rigid airfield pavement systems. The capability of AI-based surrogate-response models discussed in this paper for analyzing and designing rigid airfield pavement systems was successfully demonstrated.

Acknowledgements The authors wish to express their sincere appreciation to the Federal Aviation Administration (FAA) for supporting this study. They also express their appreciation to including, but not to limit to, Dr. David R. Brill at FAA, Dr. Kasthurirangan Gopalakrishnan, Dr. Adel Rezaei-Tarahomi, and other research team members from the ISU Program for Sustainable Pavement Engineering and Research (PROSPER) at the InTrans for their assistance in this research study.

References

1. Kaya O, Ceylan H, Kim S, Rezaei-Tarahomi A (2022) Evaluation of the Federal Aviation Administration's rigid airfield pavement cracking failure models. J Transp Eng Part B Pavements 148(1):04021071
2. Kaya O (2019) Use of soft computing and numerical analysis in design, analysis and management of pavement systems. Doctoral dissertation, Iowa State University
3. Ceylan H, Tutumluer E, Barenberg EJ (1999) Artificial neural networks for analyzing concrete airfield pavements serving the Boeing B-777 aircraft. Transp Res Rec 1684(1):110–117
4. Kaya O, Ceylan H, Kim S, Waid D, Moore BP (2020) Statistics and artificial intelligence-based pavement performance and remaining service life prediction models for flexible and composite pavement systems. Transp Res Rec 2674(10):448–460
5. Kaya O (2015) Investigation of AASHTOWare Pavement ME design/Darwin-ME TM performance prediction models for Iowa pavement analysis and design. Master's thesis, Iowa State University
6. Rezaei-Tarahomi A, Kaya O, Ceylan H, Gopalakrishnan K, Kim S, Brill DR (2022) ANNFAA: artificial neural network-based tool for the analysis of Federal Aviation Administration's rigid pavement systems. Int J Pavement Eng 23(2):400–413

7. Rezaei-Tarahomi A, Ceylan H, Gopalakrishnan K, Kim S, Kaya O (2019) Artificial neural network models for airport rigid pavement top-down critical stress predictions: sensitivity evaluation. In: Airfield and highway pavements 2019: innovation and sustainability in highway and airfield pavement technology, 2019. American Society of Civil Engineers, Reston, Virginia, USA
8. Rezaei-Tarahomi A, Kaya O, Ceylan H, Gopalakrishnan K, Kim S, Brill DR (2017) Neural networks based prediction of critical responses related to top-down and bottom-up cracking in airfield concrete pavements. In: Tenth international conference on the bearing capacity of roads, railways and airfields
9. Rezaei-Tarahomi A (2019) Development of rigid airfield pavement foundation response and moduli prediction models. Iowa State University
10. Rezaei-Tarahomi A, Kaya O, Ceylan H, Kim S, Gopalakrishnan K, Brill DR (2017) Development of rapid three-dimensional finite-element based rigid airfield pavement foundation response and moduli prediction models. Transp Geotech 13:81–91

Shear and Hydraulic Behavior of Compacted Foamed Glass Aggregates for Transportation Earthworks

Jennifer E. Nicks, Ismaail I. Ghaaowd, and Michael T. Adams

Abstract Recycled foamed glass aggregate (FGA) is an emerging material in the USA for use as a lightweight structural backfill. FGA is considered a sustainable solution, with advantageous engineering properties noted throughout the literature, but since the FGA market in the USA began less than 10 yr ago, independently assessing the behavior of the FGAs available for highway transportation agencies became necessary. The Federal Highway Administration (FHWA) has therefore made a concerted effort to evaluate the physical, mechanical, and hydraulic properties of US-manufactured, recycled FGAs under varying compaction efforts and loading conditions. This paper presents some initial findings from large-scale laboratory testing performed to evaluate shear strength and permeability of FGA under different compaction ratios and applied normal stresses. The overall results demonstrate that the breakage of FGAs due to compaction can have advantageous impacts, such as an increase in density and shear strength. While compaction does reduce permeability, FGA is still a free-draining material. The findings suggest that FGA is a viable substitute for sustainable, niche geotechnical applications; however, understanding the in-place FGA engineering properties is needed to develop better construction specifications and design guidance to ensure serviceability in transportation earthwork projects.

Keywords Foamed glass aggregate · Cellular glass · Foamed glass gravel · Creep · Long-term settlement · Shear strength · Backfill · Lightweight aggregate

J. E. Nicks (✉) · M. T. Adams
Federal Highway Administration, Turner-Fairbank Highway Research Center, McLean, VA 22101, USA
e-mail: jennifer.nicks@dot.gov

M. T. Adams
e-mail: mike.adams@dot.gov

I. I. Ghaaowd
Highway Technology Partners, Turner-Fairbank Highway Research Center, McLean, VA 22101, USA
e-mail: ismaail.ghaaowd.ctr@dot.gov

B. Indraratna and C. Rujikiatkamjorn (eds.), *Recent Advances and Innovative Developments in Transportation Geotechnics*,
https://doi.org/10.1007/978-981-97-8245-1_7

1 Introduction

Foamed glass aggregate (FGA), also known as foamed glass gravel and cellular glass, is a lightweight geomaterial used in retaining walls, bridge abutments, and embankments; as unbound materials for pavements; as covers and beddings for pipes and culverts; and as a ground improvement solution in areas of highly compressible soils [1]. Manufacturers create FGA by mixing a powder of crushed, recycled glass with a foaming agent and then conveying the mixture through a kiln at temperatures typically up to 900 °C [2–4]. When cooled, the product breaks apart into aggregate-like particles that are similar in appearance to natural pumice. Each FGA manufacturer has its own proprietary blend of glass powder, foaming agent, other additives, and water, as well as variations in kiln temperature and cook time, all of which can affect the material properties of FGA [2, 3, 5].

Considered a sustainable material, FGA is increasingly used in transportation earthworks due to its low unit weight, excellent drainage properties, and high friction angle. However, FGA is a lightweight, brittle material that is buoyant and can easily be broken apart by hand. As such, appropriate site selection and overburden are key in design, particularly when using the material to facilitate drainage or if it will be inundated with water [6]. In addition, overcompaction during placement should be avoided, depending on the design requirements, to limit crushing and, thus, increase the density of the material [7]. Typical FGA lift reductions due to compaction have been found to range between 7.6 and 16.6% depending on the method specification and equipment [8].

Studies have shown the impact of compaction on FGA breakage and behavior as it relates to thermal conductivity [9], deformation under different applied loads [10–13], and shear strength [12]; however, the influence of compaction on permeability has not been reported. Actual permeability test data reported in the literature are minimal. Based on manufacturers' results, permeability appears to range quite widely, from 864 [7] to 3800 m/d [14], up to 8640 m/d [6, 15].

FGA has been widely used in the European market [3], yet production in the US market only began in 2017 [16]. For more widespread use, particularly in the USA, additional research is needed to establish confidence in the behavior and performance of FGA. The Federal Highway Administration (FHWA) has, therefore, made a concerted effort to evaluate the physical, mechanical, and hydraulic properties of a common US-manufactured FGA. This paper presents the initial results of a study to evaluate the impact of compaction on shear strength and permeability of FGA.

2 Materials and Methods

2.1 Materials

The FGA evaluated in this study was manufactured in the northeast USA from 100% post-consumer recycled glass mixed with silicon carbide in a dry foaming process, referred to herein as FGA-SiC. Per ASTM D448-12, FGA-SiC is classified as a No. 3 coarse aggregate [17], and per ASTM D2487, a poorly graded gravel (GP) [18]. The initial gradations of FGA-SiC are shown in Fig. 1 after 10-sieve analyses were performed to understand particle size variability. The largest variations are seen in the coarse and finer particle sizes; however, coefficients of variation (COVs) for particle sizes ranged from about 3.0–7.5%, indicating that this material was consistently supplied.

The minimum ($\rho_{d,min}$) and maximum ($\rho_{d,max}$) dry densities of FGA-SiC were determined per Method A of ASTM D4254-16 [19] and per Method 1A of ASTM D4253-16 [20], respectively; $\rho_{d,min}$ was 0.23 g/cm^3, and $\rho_{d,max}$ was 0.30 g/cm^3. ASTM D4253-16 acknowledges that the test does not necessarily obtain the absolute $\rho_{d,max}$ [20]. This is especially true for FGA wherein additional vibration could result in differing amounts of breakage and, thus, different $\rho_{d,max}$ values; however, Nicks et al. showed that the test produces reasonable results when compared to a conventional backfill, even considering the brittle nature of the FGA [10].

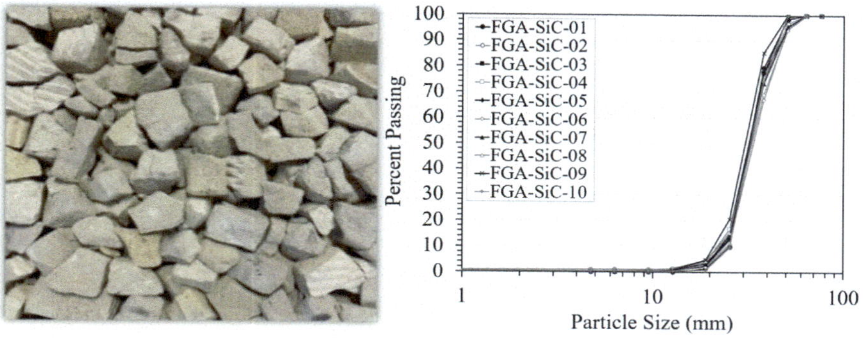

a . b .

Fig. 1 **a** FGA-SiC material and **b** initial gradations. *Source* FHWA

2.2 Methods

Shear Strength. One of the primary design parameters for aggregates used in transportation earthworks is shear strength, namely the friction angle. Therefore, large-scale direct shear (LSDS) tests were performed per ASTM D3080-11 [21]; the nominal shear box dimensions were 304.8 × 304.8 × 203.2 mm. Specimens were placed and compacted in one lift within the shear box at compaction ratios (CRs) of 0% (i.e., loose), 11%, and 22%. Note that CR is defined herein as the percent compaction from the initial, loose placement height. The range selected encompasses the typical FGA lift thickness reductions in the field due to compaction [7]. Compaction was performed on top of the specimen using a vibratory compactor with 8.8 N m of impact energy.

LSDS tests were performed under dry conditions at applied normal stress levels of 34.4, 82.7, and 206.8 kPa. Nicks et al. found negligible difference between dry and saturated test conditions for FGA [10]. All specimens were sheared at a constant strain rate of 0.381 mm/min after a 10-min consolidation phase. Before shearing, a gap was created between the upper and lower shear boxes. The gap size was nominally based on the specimens' initial D_{85} (i.e., particle size in which 85% of particles are finer) [22]; however, because the particle size of FGA-SiC would result in a gap size larger than could be accommodated by the LSDS device, a parallel gradation approach was selected to determine an appropriate, approximately 27-mm gap size based on a reduced D_{85} [10].

Permeability. Constant head tests were conducted per AASHTO T 215-23 [23] on a specimen 150 mm in diameter by 116 mm in length. The material was placed/compacted in the permeameter in one lift to the desired CR (i.e., 0, 11, and 22%), with three replicate samples prepared per CR. Tests were performed for each sample under multiple hydraulic gradients to determine the saturated coefficient of permeability (k_{sat}).

3 Results

3.1 Shear Strength

The shear stress–deformation results are presented in Table 1.

The results indicate that increased compaction effort led to higher τ_p; the difference between 22 and 0% CR is approximately 35% under 34.4 kPa normal stress and 20% under 206.8 kPa. In addition, compaction resulted in smaller horizontal strains required to mobilize $\phi_{s,p}$ (Fig. 2), which is similar to that found by Swan et al. [12]. Of note, ϕ_r after 20% lateral strain is nearly equal to $\phi_{s,p}$ for FGA-SiC. This finding may indicate that FGA could be more advantageous than conventional backfills in applications where large deformations are expected.

Table 1 LSDS test results

Target normal stress (kPa)	CR (%)	τ_p (kPa)	$\phi_{s,p}$ (°)	ϕ_r (°)	ψ_{max} (°)
34.4	0	83.8	67.7	67.1	7.2
	11	103.3	71.6	67.4	15.2
	22	128.5	75.0	72.9	17.9
82.7	0	121.9	55.9	55.4	4.8
	11	148.7	60.9	56.6	5.3
	22	167.3	63.7	60.6	7.5
206.8	0	215.2	46.2	45.8	0
	11	261.2	51.7	50.1	0
	22	267.5	52.3	49.2	0

Note τ_p = peak shear stress; $\phi_{s,p}$ = peak friction angle; ϕ_r = residual friction angle; ψ_{max} = maximum dilation angle

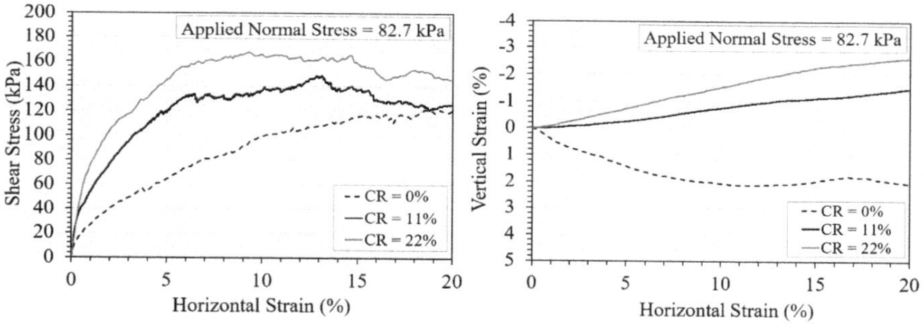

Fig. 2 Shear stress–deformation behavior of FGA-SiC. *Source* FHWA

Another aspect found through LSDS testing is the dilative behavior of FGA (Fig. 2). As CR increased, the maximum dilation angle (ψ_{max}) also increased. At the highest applied normal stress, however, the samples experienced compression throughout shear regardless of CR, indicating that a considerable contribution of shear strength was due to breakage versus dilation.

The resulting Mohr–Coulomb failure envelopes are shown in Fig. 3. Tangent friction angles were 37.3°, 42.4°, and 38.9° for CRs of 0%, 11%, and 22%, respectively. Due to the crushing and interlock of the aggregates, apparent cohesion is also present and increases as CR gets larger (Fig. 3).

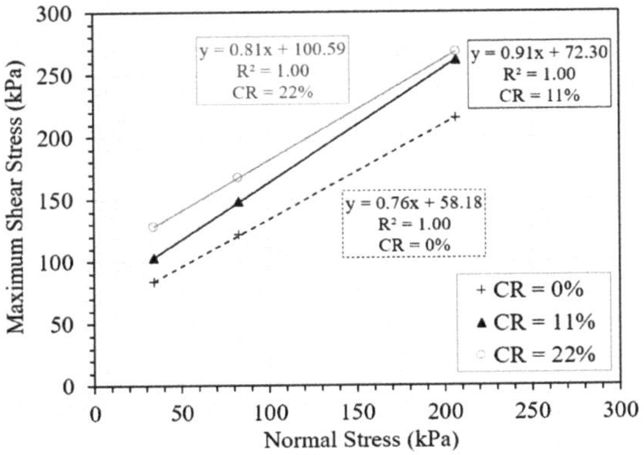

Fig. 3 Linear Mohr–Coulomb failure envelopes for FGA-SiC. *Source* FHWA

3.2 Permeability

The results from permeability testing are shown in Table 2.

As expected, increased CR and thus decreased porosity resulted in decreased permeability. Regardless, even at the highest CR tested (i.e., 22%), excellent drainage values were still found. The findings indicate a linear relationship between k_{sat} and CR, with k_{sat} linearly decreasing with increasing CR (Fig. 4). The COVs are remarkedly low for a geomaterial and decrease with increasing CR (Table 2). This suggests that the variability in FGA drainage behavior reduces with higher compactive effort; a finding also discovered as it relates to static compression [9]. Additional tests at higher CRs would help reinforce this finding.

Table 2 Constant head permeability test results

CR	ρ_d (g/m^3)	Porosity	k_{sat} (m/d)	μ_k (m/d)	σ_k (m/d)	COV (%)
0	0.24	0.54	13,168	14,399	1188	8.25
			14,025			
			16,004			
11	0.27	0.48	13,574	12,488	903	7.23
			12,526			
			11,364			
22	0.30	0.43	8542	8923	279	3.13
			9023			
			9204			

Note ρ_d = dry density; μ_k = mean k_{sat}; σ_k = standard deviation of k_{sat}

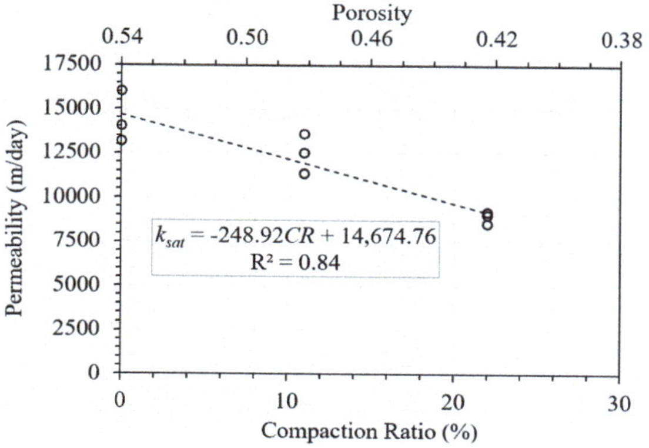

Fig. 4 Coefficient of saturated permeability as a function of CR. *Source* FHWA

3.3 Particle Breakage

The initial and post-test gradations were measured to determine the amount of FGA breakage due to sample preparation, compaction, and testing. Permeability tests provided post-CR breakage data, while shear tests provided post-CR and shear breakage data; example pre- and post-test sieve results are shown in Fig. 5a.

For the loose condition (i.e., CR = 0%), simple placement of the material still resulted in some breakage, albeit negligible; however, after the sample sheared, the breakage index [24] ranged from about 9–30%, linearly increasing with increasing applied normal stress (Fig. 5b). The results indicate that the role of the applied normal stress on breakage diminishes with increasing CR. For example, at 34.4 kPa of applied normal stress, sample preparation caused about half of the total breakage under loose conditions, yet at a CR of 22%, that same applied stress did not result in

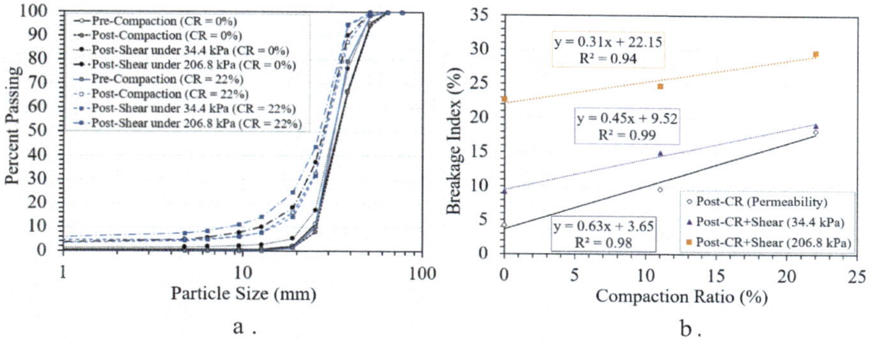

Fig. 5 **a** FGA gradations pre- and post-testing and **b** breakage index as a function of CR. *Source* FHWA

any appreciable increase in breakage. Note that the impact of compaction on internal micro-fractures and air void content within the aggregate itself was not investigated in this stage of the work.

4 Conclusions

The results of this study suggest that compaction of FGA is of little concern with respect to shear strength and permeability, provided that the increase in density is tolerable for the application. While compaction increased shear strength, it decreased permeability; however, FGA remained a free-draining material even under the highest compactive efforts expected in the field. The findings suggest that FGA is a viable substitute for sustainable geotechnical applications; however, understanding its in-place engineering properties is needed to develop better construction specifications and design guidance to ensure serviceability in transportation earthwork projects. Future efforts include looking at additional engineering properties, including freeze–thaw durability and cyclic behavior for compaction ratios feasible within current practice, focusing on evaluating different types of US-manufactured FGA.

References

1. Qamhia IIA, Tutumluer E, Nicks JE, Adams MT, Khan MS (2023) Lightweight and alternative backfills for highway applications: state-of-the-art practice in the U.S.A. Transp Res Rec. https://doi.org/10.1177/03611981231191519. Last accessed 6 Feb 2024
2. El-Haggar SM (2021) Sustainability of municipal solid waste management. In: El-Haggar SM (ed) Sustainable industrial design and waste management. Academic Press, pp149–196. https://doi.org/10.1016/B978-012373623-9/50007-1. Last accessed 6 Feb 2024
3. Hibbert M (2016) Understanding the production and use of foam glass gravel across Europe and opportunities in the UK, Final Report. CIWM, Northampton, UK
4. Zegowitz A (2010) Cellular glass aggregate serving as thermal insulation and a drainage layer. Paper presented at the Building XI Conference, American Society of Heating, Refrigerating and Air-Conditioning Engineers, Clearwater Beach, FL
5. Spiridonov YA, Orlova LA (2003) Problems of foam glass production. Glass Ceram 60(9–10):313–314. https://doi.org/10.1023/B:GLAC.0000008234.79970.2c. Last accessed 6 Feb 2024
6. Loux TA, Filshill A (2021) Foamed glass aggregate for resilient waterfront construction. Geotechnical Special Publication No. 328, Geo-Extreme, ASCE, Reston, VA
7. Loux TA, Filshill A, Zhang Z (2019) Foamed glass aggregate lightweight fill over compressible soils. In: Proceedings of the 72nd Canadian geotechnical conference, St. Johns, Newfoundland
8. McGuire MP, Loux TA, VandenBerge DR (2021) Field-scale tests to evaluate foamed glass aggregate compaction. In: Proceedings of the international foundations congress and equipment expo, pp 157–168. American Society of Civil Engineers, Dallas, TX. https://doi.org/10.1061/9780784483435.015. Last accessed 6 Feb 2024
9. Mustafa WS, Szendefy J, Nagy B (2023) Thermal performance of foam glass aggregate at different compaction ratios. Buildings 13(7). https://doi.org/10.3390/buildings13071844. Last accessed 6 Feb 2024

10. Nicks JE, Ghaaowd II, Adams MT (2023) Strength-deformation characteristics of two different foamed glass aggregates under a static design load. ASTM Geotech Test J 47(1)
11. Arulrajah A, Disfani MM, Maghoolpilehrood F, Horpibulsuk S, Udonchai A, Imteaz M, Du Y-J (2015) Engineering and environmental properties of foamed recycled glass as a lightweight engineering material. J Clean Prod 94(1):369–375. https://doi.org/10.1016/j.jclepro.2015.01.080. Last accessed 6 Feb 2024
12. Swan RH, Yeom S, Sjoblom KJ, Stark TD, Filshill A (2018) Engineering properties of foamed recycled glass as a lightweight fill. In: Proceedings of geo-Chicago 2016: sustainable materials and resource conservation, Geotechnical Special Publication 272. American Society of Civil Engineers, Chicago, IL, pp 11–22
13. Mustafa WS, Nagy B, Szendefy J (2022) Impact of compaction ratio and loading period on compressional behavior of foam glass aggregates. Constr Build Mater 343(1):1–11. https://doi.org/10.1016/j.conbuildmat.2022.128111. Last accessed 6 Feb 2024
14. GEOCELL (2022) GEOCELL foam glass gravel: high performance in every aspect
15. Foamit (2016) Foamed glass applications. Forssa, Finland. https://www.foamit.fi/wp-content/uploads/2016/10/Foamedglass.pdf. Last accessed 6 Feb 2024
16. Gibson S (2019) Two startups seek to grow interest in foam glass fill. Green Building Advisor, Taunton Press, Newtown, CT. http://web.achive.org/web/20220927115342 https://www.greenbuildingadvisor.com/article/two-startups-seek-grow-interest-foam-glass-fill. Last accessed 6 Feb 2024
17. ASTM International (2017) Standard specification for sizes of aggregate for road and bridge construction. ASTM D448-12. ASTM International, West Conshohocken, PA, approved 1 Oct 2017. https://doi.org/10.1520/D0448-12R17. Last accessed 6 Feb 2024
18. ASTM International (2017) Standard practice for classification of soils for engineering purposes (unified soil classification system). ASTM D2487-17. ASTM International, West Conshohocken, PA, approved 15 Dec 2017. https://doi.org/10.1520/D2487-17E01. Last accessed 6 Feb 2024
19. ASTM International (2016) Standard test methods for minimum index density and unit weight of soils and calculation of relative density. ASTM D4254. ASTM International, West Conshohocken, PA, approved 1 March 2016. https://doi.org/10.1520/D4254-16. Last accessed 6 Feb 2024
20. ASTM International (2016) Standard test methods for maximum index density and unit weight of soils using a vibratory table. ASTM D4253. ASTM International, West Conshohocken, PA, approved 1 March 2016. https://doi.org/10.1520/D4253-16E01. Last accessed 8 Feb 2024.
21. ASTM International (2011) Standard test method for direct shear test of soils under consolidated drained conditions. ASTM D3080/D3080M. ASTM International, West Conshohocken, PA, approved 1 Nov 2011. https://doi.org/10.1520/D3080_D3080M-11. Last accessed 8 Feb 2024
22. Nicks JE, Gebrenegus G, Adams MT (2015) Strength characterization of open-graded aggregates for structural backfills. FHWA Report No. HRT-15-034. Federal Highway Administration, Washington, DC
23. AASHTO (American Association of State Highway and Transportation Officials) (2023) standard method of test for permeability of granular soils (constant head). AASHTO T 215 (2023). AASHTO, Washington, DC, approved 2023
24. Marsal RJ (2024) Large-scale testing of rockfill materials. J Soil Mech Found Div 93(2):27–43. https://doi.org/10.1061/JSFEAQ.0000958. Last accessed 6 Feb 2024

Unsaturated Soil Mechanics in Engineering Practice Using the Equivalent Method

Adrian R. Russell

Abstract The objective of this paper is to assist the uptake of unsaturated soil mechanics in strength and stability problems in engineering practice. It gives examples of research outcomes related to slope and shallow foundation stability determinations and shows how they can be applied routinely in hand calculations and commercial software using the equivalent method. The equivalent method involves capturing suction influences through an equivalent cohesion and an equivalent unit weight and then doing a total stress analysis. It also outlines industry changes and trends, including good practices and commonly made mistakes, relevant to way unsaturated soils are accounted for and the way suction is measured and implemented into a geotechnical model.

Keywords Unsaturated soil · Suction · Effective stress · Strength · Stability

1 Introduction

Unsaturated soils are widespread and need to be dealt with in many engineering problems and infrastructure types. They are mixtures of solid particles plus air and water which occupy the pore space. The coexistence of air and water gives rise to suction through capillary action. The suction increases particle-to-particle contract forces and elevates the soil's effective stress, strength and stiffness.

A. R. Russell (✉)
Centre for Infrastructure Engineering and Safety, School of Civil and Environmental Engineering, UNSW, Sydney, NSW 2052, Australia
e-mail: a.russell@unsw.edu.au

B. Indraratna and C. Rujikiatkamjorn (eds.), *Recent Advances and Innovative Developments in Transportation Geotechnics*,
https://doi.org/10.1007/978-981-97-8245-1_8

For many decades, there has been a lot of research which addresses the fundamental engineering behaviour of unsaturated soils and techniques to measure key properties. There have been considerable efforts to develop constitutive models describing the stress–strain characteristics, explanations of water retention characteristics, and the development of new experimental testing and field monitoring devices. The literature is extensive and has been summarised in a recent state-of-the-art report [1].

Despite these advances, the use of unsaturated soil mechanics in engineering practice is not routine. The author has seen considerable industry uptake growth in recent years, however. This is partly due to government agencies encouraging the use of unsaturated soil mechanics in infrastructure designs, for example in South Australia, Australia [2]. While that may be localised to one state in one country, it is validation and acknowledgement by a major infrastructure owner that unsaturated soil mechanics can lead to efficiencies and added reliability. It may motivate others to embrace it as well. Also, several of the world's largest mining companies have realised there are benefits to be gained by using unsaturated soil mechanics when assessing the safety and stability of their tailings storages and filtered tailings stacks.

Here a simple technique is presented that enables unsaturated soil mechanics to be accounted for in strength and stability problems. It is sufficiently accurate from a practical viewpoint. It is called the equivalent method because central to it is the quantification of an equivalent cohesion and an equivalent unit weight. Those quantities are then used in a total stress analysis. The idea of equivalency when dealing with the cohesion part of an unsaturated soil's strength is already well known. What is not well known is that a companion equivalent unit weight also needs to be considered. This paper presents the complete approach. It is explained under what circumstances the equivalent method applies, it being when the unsaturated soils behave as though they are pseudo-drained.

2 Interrelation Between Water Retention and Effective Stress

A suction s is present in an unsaturated soil and causes the particle contact forces to increase compared to when saturated or dry. The macroscopic effects are increases in the effective stress and strength. The effective stress σ' is defined as [3, 4]

$$\sigma' = \sigma - u_a + \chi s \tag{1}$$

where σ is total stress, χ is the effective stress parameter, u_{a} and u_{w} are the pore air and pore water pressures, respectively, $s = u_{\mathrm{a}} - u_{\mathrm{w}}$ is the suction, χs is the contribution of suction to the effective stress and $\chi s - u_{\mathrm{a}}$ is the combined contributions of pore air and pore water pressures to the effective stress. Here atmospheric pressure is used as the datum, meaning σ, u_a or u_w are zero when they are equal to that. $\chi = 0$ for dry soils so $\sigma' = \sigma - u_a$. $\chi = 1$ for saturated soils so $\sigma' = \sigma - u_w$.

To establish χ it is necessary to consider how water is retained in the soil. The relationship between degree of saturation S_r and s is relevant, established usually through experimentation, and is referred to as a water retention curve (WRC). The relationship between s and S_r depends on void ratio e as well.

The WRC comprises a main drying curve and a main wetting curve. For a given e and s, the S_r on the main drying curve is larger than that on the main wetting curve. When a soil experiences a change from drying to wetting, or vice versa, the hydraulic state scans between the main drying and wetting curves on what are known as scanning curves. There are an infinite number of scanning curves. A single value of S_r corresponds to an infinite number of possible s values for any given e and the exact value depends on the mechanical loading and hydraulic history.

For a given e, it is often observed that S_r relates to s in a power laws [5, 6]. The main wetting and drying curves and scanning curves are often idealised as being linear in the double logarithmic S_r versus s plane (Fig. 1). S_r on the main drying and wetting curves may be defined as

$$S_r = \begin{cases} 1 & \text{for } s \le s_e \\ \left(\frac{s}{s_e}\right)^{\alpha} & \text{for } s \ge s_e \end{cases} \tag{2}$$

where α is the slope of the curves. s_e is the s value separating saturated from unsaturated states. For the main drying curve $s_e = s_{ae}$, where s_{ae} is the air entry suction. For the main wetting curve $s_e = s_{ex}$, where s_{ex} is the air expulsion suction. On scanning curves, S_r is defined as

$$S_r = \begin{cases} \left(\frac{s_{rd}}{s_{ae}}\right)^{\alpha}\left(\frac{s}{s_{rd}}\right)^{\beta} & \text{for drying path reversal } \; s \le s_{rd} \\ \left(\frac{s_{rw}}{s_{ex}}\right)^{\alpha}\left(\frac{s}{s_{rw}}\right)^{\beta} & \text{for wetting path reversal } \; s_{rw} \le s \end{cases} \tag{3}$$

where β is the slope of the scanning curves. s_{rd} and s_{rw} represent intercepts of the scanning curves with the main drying and main wetting curves, respectively, and also represent points of suction reversal on the main drying and main wetting curves if suction is suddenly reduced or increased, respectively.

The WRC for a certain e can be made applicable to any other e through s_e. When the particle size distribution is fractal, like with most residual soils (e.g. decomposed granite), loess and crushed/ground sands, gravels, cobbles, boulders (e.g. tailings, ore, glacial till), expressions for s_{ae} and s_{ex} are [6]

$$s_{ae} = C_1 e^{-D_s} \quad \text{and} \quad s_{ex} = C_2 s_{ae} \tag{4}$$

in which C_1 is a positive constant with units of stress, D_s is the fractal dimension of the particle size distribution and C_2 is a dimensionless constant [6].

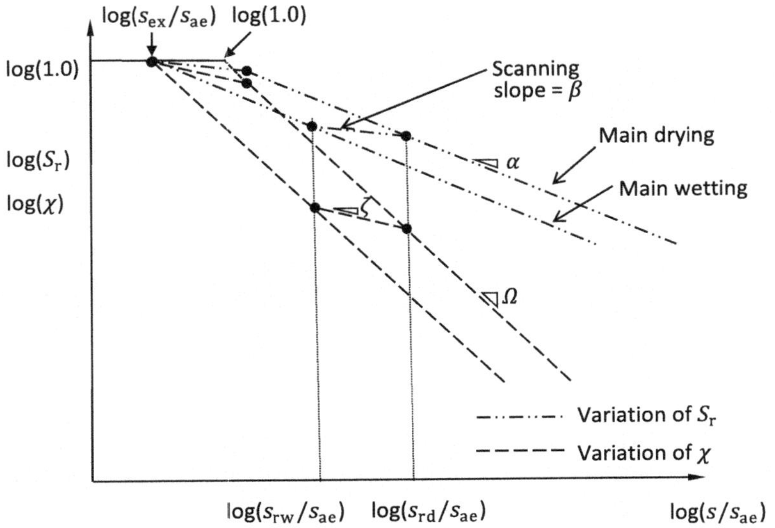

Fig. 1 S_r and χ plotted against s/s_{ae} in double logarithmic planes

Power laws also relate χ to s/s_e [7]. When the hydraulic state is located on the main wetting or main drying curve

$$\chi = \begin{cases} 1 & \text{for } \frac{s}{s_e} \leq 1 \\ \left(\frac{s}{s_e}\right)^{\Omega} & \text{for } \frac{s}{s_e} \geq 1 \end{cases} \tag{5}$$

and when on the scanning curve

$$\chi = \begin{cases} \left(\frac{s_{rd}}{s_{ae}}\right)^{\Omega}\left(\frac{s}{s_{rd}}\right)^{\zeta} & \text{for drying path reversal} \quad s \leq s_{rd} \\ \left(\frac{s_{rw}}{s_{ex}}\right)^{\Omega}\left(\frac{s}{s_{rw}}\right)^{\zeta} & \text{for wetting path reversal } s_{rw} \leq s \end{cases} \tag{6}$$

Ω and ζ are the slopes of curves when hydraulic states are on main drying/wetting curves and scanning curves, respectively. Ω has a best-fit value of -0.55 for a wide range of soil types [7]. Here it is taken that $\frac{\Omega}{\alpha} = \frac{\zeta}{\beta}$ for all hydraulic state locations to ensure compatibility. The relationship between χ and s is also shown in Fig. 1. More simply, χ can be derived directly from S_r, for any location of the hydraulic state, using

$$\chi = S_r^{\Omega/\alpha} = S_r^{\zeta/\beta} \tag{7}$$

An expression for χs can be obtained using the above definitions, in terms of S_r or alternate and commonly used soil properties. For example, in terms of gravimetric water content w, specific gravity G_s and e, noting $wG_s = S_r e$, the expression is

$$\chi s = C(G_s w)^{(1+\Omega)/\alpha} e^{-D_s-(1+\Omega)/\alpha} \tag{8}$$

in which $C = C_1$ if the hydraulic state is on the main drying curve, $C = C_1 C_2$ if on the main wetting curve and $C = (C_1 \times C_1 C_2)^{1/2}$ if at midpoint of a scanning curve.

3 Incorporating Suction into Strength and Stability Problems Using the Equivalent Method

The strength and stability of shallow foundations and curvilinear slopes in unsaturated soil, in two dimensions, were studied by [8, 9] using slip line theory. The unsaturated soil is homogenous and perfectly plastic, obeys the Mohr–Coulomb failure criterion, the friction angle φ' is constant and the χs profile and cohesion c' vary linearly with depth. The shear strength τ is given by

$$\tau = c' + \sigma' \tan\varphi' \quad \text{or} \quad \tau = c' + \chi s \tan\varphi' + \sigma \tan\varphi' \tag{9}$$

In Eq. 9, and hereafter, $u_a = 0$ is assumed. Also, c' and φ' are assumed to be independent of soil density, confining pressure and loading history for simplicity, despite small dependencies. The slight elevation of φ' which occurs due to suction hardening is also ignored for simplicity.

The χs and c' profiles obey

$$\chi s = (\chi s)_0 + K_{\chi s} z \quad \text{and} \quad c' = c'_0 + K_c z \tag{10}$$

in which $(\chi s)_0$ and c'_0 are the values at the ground surface, z is the depth (zero at the ground surface, becoming increasingly positive as depth increases) and $K_{\chi s}$ and K_c are the rates of change.

The problem configuration for the shallow foundation is shown in Fig. 2. Examples of slip lines around a smooth and rough strip footing are in [8].

The three-dimensionless groups which control bearing capacity are V, F and φ', with V and F defined as

$$V = \frac{q_u}{c'_0 + (\chi s)_0 \tan\varphi' + q \tan\varphi'} \quad \text{and} \quad F = \frac{B(\gamma_t + K_c \cot\varphi' + K_{\chi s})}{q + (\chi s)_0 + c'_0 \cot\varphi'} \tag{11}$$

in which B is the footing width, q_u is the bearing capacity, q is a surcharge on the soil surface adjacent to the footing and γ_t is the total unit weight. Vo and Russell [8]

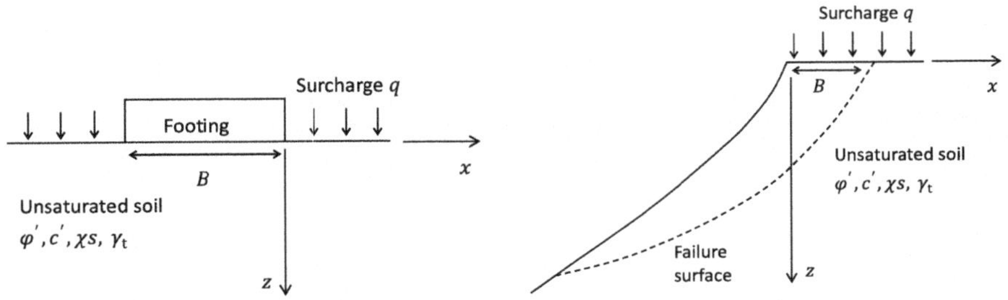

Fig. 2 Problem configurations for a shallow foundation and slope

demonstrate the relationships between V, F and the well-known bearing capacity factors and equation.

The problem configuration for the slope is also shown in Fig. 2. Examples of slip lines are in [9].

The three-dimensionless groups which control slope failure are T, F and φ', with F defined exactly as for the shallow foundation case, with B representing the distance between the top of the slope and where the failure surface intercepts the ground surface, q is a surcharge on the soil surface, and T is

$$T = \frac{\cot\varphi'}{2}\ln\left|\frac{((\chi s)_0 + c_0'\cot\varphi')(1+\sin\varphi')}{(q + (\chi s)_0 + c_0'\cot\varphi')(1-\sin\varphi')}\right| \tag{12}$$

Any tension cracked soil is treated as part of the surcharge and is not shown in the problem geometries in Fig. 2.

These studies reveal a way that s influences may be accounted for. In limit equilibrium software, or hand analysis, s (and the variation of c) may be captured through two terms referred to here as an equivalent unit weight and an equivalent cohesion

$$\gamma_{eq} = \gamma_t + K_c\cot\varphi' + K_{\chi s} \quad \text{and} \quad c_{eq} = c_0' + (\chi s)_0\tan\varphi' \tag{13}$$

A total stress analysis may then be adopted when cohesion and unit weight are set to be equal to c_{eq} and γ_{eq}. No other influences of s need to be accounted for.

This ensures the dimensionless groups which control strength and stability, and the governing mechanics, are preserved and the stability computed is representative of the real problem. The dimensionless groups V, F and T defined above may be rewritten as

$$V = \frac{q_u}{q\tan\varphi' + c_{eq}},\quad F = \frac{B\gamma_{eq}}{q + c_{eq}\cot\varphi'},$$
$$\text{and}\quad T = \frac{\cot\varphi'}{2}\ln\left|\frac{c_{eq}\cot\varphi'(1+\sin\varphi')}{(q + c_{eq}\cot\varphi')(1-\sin\varphi')}\right|. \tag{14}$$

It may be wise to ensure the χs profile adopted is representative of a worst case, involving a wetting process in some way.

This approach circumvents the need to use the inbuilt unsaturated soil features of a software, as long as the χs profile can be assumed (nearly) constant as failure is approached, i.e. when the soil is pseudo-drained. Vo et al. [10] and Russell et al. 11] argued that pseudo-drainage applies if $v_a/v > 0.15$ for slow and fast loading situations, and for any v_a/v for slow loading situations. v_a/v is the volumetric air ratio in the soil and is defined as $\frac{v_a}{v} = \frac{e}{1+e}(1 - S_r)$, where the volume of air is v_a and the overall volume is v.

Applying the equivalent method to unsaturated soils, especially the altered definition of the unit weight, is similar to what is done routinely for saturated soils. To demonstrate the similarity suppose the soil is saturated all the way to the ground surface and hydrostatic conditions apply. This means $(\chi s)_0$ is zero and $K_{\chi s} = \partial(-u_w)/\partial z = -\gamma_w$. Also suppose the soil is cohesionless so c_0', K_c and c_{eq} are zero. It follows that $\gamma_{eq} = \gamma_t + K_c \cot\varphi' + K_{\chi s} = \gamma_t - \gamma_w = \gamma'$, that is the effective unit weight, and is what is used in stability calculations when drained conditions prevail.

4 Example of Application to a Slope Stability Problem

A hypothetical slope comprising unsaturated soil, shown in Fig. 3, is known to have failed, with failure to correspond to a factor of safety of 1. The failure surface is also shown. The width of the top of the sliding mass is $B = 14.6$ m. The weight of the tension cracked soil is included in the surcharge q of 228 kPa. It is also known $\gamma_t = 18.5$ kN/m^3, $c' = 16.8$ kPa (and $K_c = 0$), $\varphi' = 30°$, $\chi s = 35.0 - 0.5z$ kPa, i.e. $(\chi s)_0 = 35.0$ kPa and $K_{\chi s} = -0.5$ kPa/m. The dimensionless groups are then $F = 0.90$, $T = -0.363$.

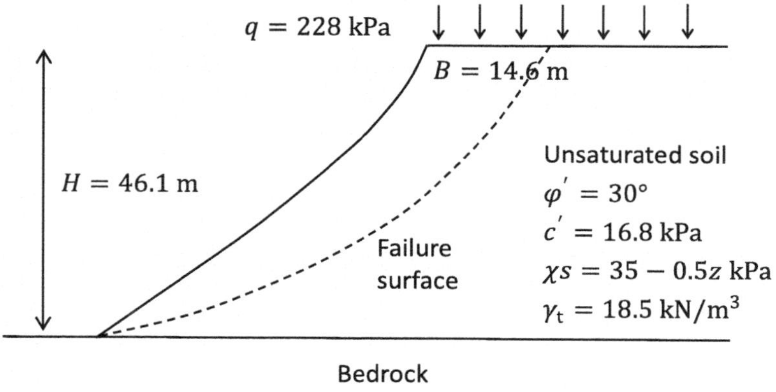

Fig. 3 An example of a failed slope

The equivalent method involves the use of $\gamma_{eq} = 18.0$ kN/m^3 and $c_{eq} = 37.0$ kPa in a total stress analysis, corresponding to the same F and T values.

An outrageous case is also considered, involving an unrealistic combination of parameters that combine to recover the same F and T values, being $\gamma_{eq} = 78.9$ kN/m^3, $c_{eq} = 162$ kPa and $q = 1000$ kPa.

The analyses for the equivalent method and the outrageous case both return a factor of safety of 1. This is expected as it is the dimensionless groups which govern the strength and stability, not the individual parameters which combine to form them.

5 Trends and Mistakes in Practice

The author, in his consulting activities and industry interactions, has observed trends and mistakes in engineering practice. They are listed below for others to learn from.

Companies are engaging university laboratories to determine WRCs. Filter paper tests and WP4C dewpoint tests are commonly used due to their speed and ease. It is good practice to use more than one technique and compare results when establishing the WRC. However, there have been cases where hydraulic hysteresis has not been considered, i.e. a failure to recognise that water retention depends on the wetting or drying history and that a main wetting curve as well as a main drying curve, and the slope of the scanning curve, need to be determined. In one case it was erroneously supposed a hydraulic state lies on the main dry curve when designing for a wetting event.

It is not uncommon for companies to use an equivalent cohesion, defined as $c' + \chi s \tan\varphi'$, in strength and stability problems. This paper demonstrates how, if a gradient of χs exists, an alternate definition of c_{eq}, as well as γ_{eq}, are to be used. Any gradient of c' must also be accounted for. Furthermore, c_{eq} and γ_{eq} should only be used when pseudo-drained conditions prevail, that is when $v_a/v > 0.15$.

Some deal with total suction, that is the sum of matric (capillary) suction (which is the suction used in this paper) and osmotic suction. The users then assume total suction contributes to strength and stiffness in a single way, irrespective of individual contributions of matric and osmotic components. The mechanics of how osmotic and matric suctions individually, and in combination, alter a soil behaviour is poorly developed, although it is generally understood that osmotic suction only has a role if clay minerals are present. In silts, sands, gravels and their mixtures it is most important to focus on matric suction. Osmotic suction will have a very minor, if any at all, influence in those. If a significant quantity of clay minerals are present, however, osmotic suction may be considered as well, but only if there is evidence that it causes a strength alteration.

Other observations include failure to realise suction needs to be scaled down using a multiplier χ before incorporating into strength. Some underappreciate the importance of material type when assessing suction influences, e.g. wrongly think a suction of 1000 kPa in a clean rock fill will provide significant strength gain. In clean and very coarse granular materials suction has a very minor role.

There have been inappropriate uses of MPS-6 (Teros 21) devices to determine suction in the field. This stems from a failure to realise they are only calibrated for when undergoing drying, and only after prolonged drying (i.e. not when drying occurs immediately after a wetting event). This is because a ceramic in the devices has its own hydraulic hysteresis. Separate calibration curves will apply to when the ceramic is experiencing main wetting or drying events, and when scanning between those.

Some users are inserting piezometers in an unsaturated soil, unaware the piezometer tip may desaturate and become unreliable. This is particularly concerning if the piezometer readings are to identify a rise in water table/phreatic surface and trigger alarms.

Many do not realise soils immediately above a phreatic surface can be saturated but have negative water pressures due to capillary rise. Those soils are incorrectly treated as though they are unsaturated. Too often the negative water pressures are neglected in determinations of effective stress when doing stability calculations or interpreting in situ test results.

6 Conclusion

The equivalent method has been outlined and enables practising engineers to incorporate suction into stability calculations of unsaturated soil slopes and shallow foundations. The calculations can be done by hand or commercial software. Suction influences are captured in an equivalent cohesion and an equivalent unit weight. Those equivalent quantities, with the effective friction angle, are then used in a total stress analysis. Equivalency of cohesion has been adopted by others already. This paper demonstrates that a companion equivalency of unit weight is needed as well.

The method only applies when pseudo-drained conditions prevail, that is when the contribution of suction to effective stress and strength does not change significantly during loading and deformation until the onset of failure. This can be expected when the unsaturated soil comprises at least 15% air by volume.

Acknowledgements The work forms part of an Australian Research Council Future Fellowship (FT200100820) awarded to the author and that funding is gratefully acknowledged.

References

1. Khalili N, Romero E, Marinho F (2022) State of the art report. Advances in unsaturated soil mechanics: constitutive modelling, experimental investigation, and field instrumentation. In: Rahman M, Jaksa M (eds) The 20th ICSMGE-state of the art and invited lectures. Australian Geomechanics Society, Sydney, Australia
2. Herraman R (2019) A lightweight soil nail retaining wall in unsaturated clay. Int J Geoeng Case Histories 5(1):1–11

3. Bishop A (1959) The principle of effective stress. Teknisk Ukeblad 106(39):859–863
4. Khalili N, Geiser F, Blight G (2004) Effective stress in unsaturated soils: review with new evidence. Int J Geomech 4(2):115–126
5. Brooks R, Corey A (1964) Hydraulic properties of porous media and their relation to drainage design. Am Soc Agric Eng Trans 7(1):26–28
6. Russell A (2014) How water retention in fractal soils depends on particle and pore sizes, shapes, volumes and surface areas. Géotechnique 64(5):379–390
7. Khalili N, Khabbaz M (1998) A unique relationship for χ for the determination of the shear strength of unsaturated soils. Géotechnique 48(5):681–687
8. Vo T, Russell A (2016) Bearing capacity of strip footing on unsaturated soils by the slip line theory. Comput Geotech 74:122–131
9. Vo T, Russell A (2017) Stability charts for curvilinear slopes in unsaturated soils. Soils Found 57:543–556
10. Vo T, Wang Y, Russell A (2022) The fall cone test in unsaturated soil and tailings pastes. Géotechnique 72(3):274–281
11. Russell A, Vo T, Ayala J, Wang Y, Reid D, Fourie A (2023) Cone penetration tests in saturated and unsaturated silty tailings. Géotechnique. https://doi.org/10.1680/jgeot.21.00261

Recommendation for the Use of Micropiles to Ensure the Reliability of the Railway Track

Askar Zhussupbekov, Abdulla Omarov, and Abilkhair Issakulov

Abstract A damaged railway line in Astana, Kazakhstan, which is located on soft soil, is the subject of an investigation, and possible remedial options are discussed in this paper. We conducted an analysis and geotechnical investigation using modern soft soil improvement techniques to identify the most effective strategies for stabilizing the damaged subgrade and to suggest remedial actions. The methods and procedures for stabilizing the soft soil were selected taking into account the need to ensure uninterrupted train traffic on the line, the largest equipment that could be used on the site, and the height restrictions imposed by overhead power lines. It was proposed to use vertical and horizontal micropiles for reinforcement to ensure the stability and reliability of the railway track. Recommendations for the restoration of the soft soil beneath the damaged section of the railway line, derived from the geotechnical investigations, are presented and illustrated in this paper. Understanding the relationship between soil and structure on problematic soft soils has been made possible by these studies.

Keywords Micropile · Railway · Georadar · Constructive measures

1 Introduction

In geotechnical engineering, the study of embankment soils beneath railroad lines is crucial. Here are a few eminent scholars and authors of relevant literature: Professor David M. Potts is renowned for his research on the behavior and soil–structure interaction of railway embankments; Professor David J. White focuses on the stability and strengthening of railway embankments; Professor J. Freeman studies the geotechnical aspects of railway systems, including soil stabilization and embankment stability; Professor Ian A. G. Smith is recognized for his work in geotechnical

A. Zhussupbekov · A. Omarov (✉) · A. Issakulov
Department of Civil Engineering, L.N. Gumilyov Eurasian National University, Astana, Kazakhstan
e-mail: omarov_01@bk.ru

B. Indraratna and C. Rujikiatkamjorn (eds.), *Recent Advances and Innovative Developments in Transportation Geotechnics*,
https://doi.org/10.1007/978-981-97-8245-1_9

engineering, which includes topics related to soil stabilization and railway embankments; Dr. John D. H. B. Gibbens studies soil mechanics and its application to railway embankments; Dr. Malcolm S. W. Casemore works on railroad embankments and soil stabilization. Research on the reinforcement of embankment soils under railway tracks has been conducted by numerous experts in the fields of geotechnical engineering and transportation infrastructure. One notable figure is Prof. John M. Burland, who has made significant contributions to soil mechanics and the reinforcement of soil structures. Additionally, Prof. R. F. Scott and Prof. A. H. Chynoweth have also conducted important research in this area. Many studies focus on different reinforcement techniques like geosynthetics, geogrids, and soil improvement methods tailored to railway embankments. Prof. M. J. Tomlinson is known for his work on soil mechanics and foundation engineering, including aspects relevant to railway embankments. Prof. A. H. Chinovet dealt with geotechnical issues related to soil reinforcement and railway embankments. Studies on the behavior of railway embankments and methods of their reinforcement were carried out by Professor S. J. T. Moore. Prof. David M. Potts examines issues related to settlement of soils, including differential settlement and its impact on infrastructure. Prof. David M. Potts work often includes the study of methods to reinforce and stabilize soils, such as the use of geosynthetics or other ground improvement techniques. "The Analysis and Design of Geotechnical Structures" covers various methods and approaches for analyzing and designing structures in geotechnical contexts, including embankments and their reinforcement. Prof. R. F. Scott is known for his work on soil mechanics, particularly in relation to soil reinforcement and foundation engineering. The paper "The Behavior of Reinforced Soil Embankments" would typically cover theoretical and practical aspects of reinforced embankments, including case studies and experimental results. Prof. A. H. Chynoweth has made significant advances in soil mechanics, foundation engineering, and embankment stability in the field of geotechnical engineering. The paper "Ground Improvement and Embankment Design" would normally go over several ground improvement methods and how to use them for embankment design [1–5].

Numerous scientists and engineers have studied this subject, concentrating on the application of pile foundations to enhance the performance and stability of railroad embankments: Dr. Andrew W. Skempton, Prof. Andrew S. R. Wang, Prof. Andrew M. Potts, Prof. R. M. H. Tatsuoka, Prof. M. J. Tomlinson et al. Prof. R. M. H. Tatsuoka has published work on the use of piles and other foundation systems to enhance the performance of embankments, including those used in railway infrastructure. The following research papers have studied various field tests and numerical methods and also considered new methods and technologies for the use of piles for infrastructure projects, including railways: The following papers are related to this topic: "Stabilization of Railway Embankments Using Pile Foundations" discusses how pile foundations are designed and used to address stability and settlement issues in railway embankments; "Performance of Pile Foundations for Railway Embankments" examines how well pile foundations reinforce embankments through field tests and case studies; "Numerical Analysis of Piled Embankments" uses numerical modeling to simulate the behavior of piled embankments under different loading conditions;

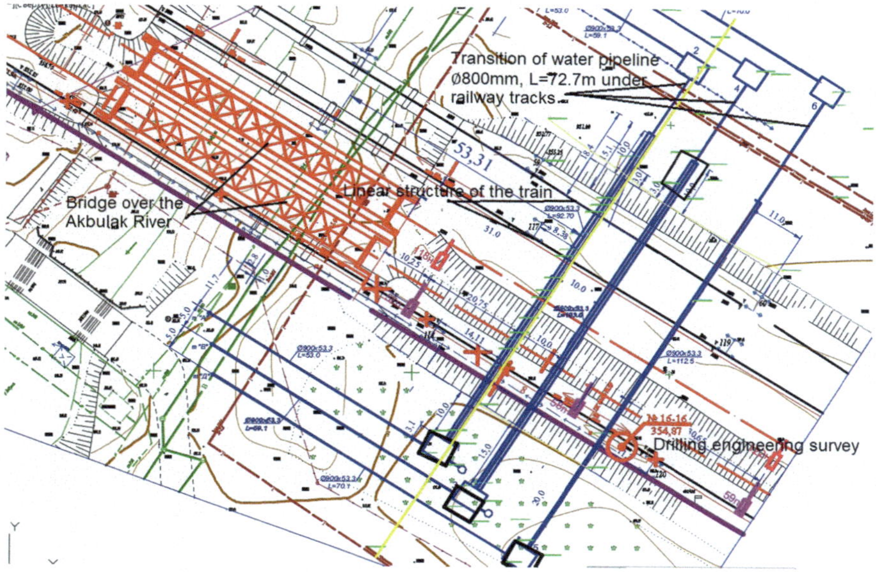

Fig. 1 Location of engineering-geological borehole

"Innovations in Pile Foundations for Transportation Infrastructure" looks at new methods and technologies for the use of piles in infrastructure projects, including railroads [5–9].

It was discovered that there was damage to a portion of the railway line in the Kazakhstani city of Astana, which is situated above the Akbulak River (Fig. 1). The existence of soft soils in the subgrade beneath the railway line at this location complicates the situation. An examination of possible restoration methods for this segment of the railway was initiated by the necessity to rectify the damage. The examination and various corrective action alternatives for the aforementioned damaged railway line section are detailed in this article (Fig. 2).

2 Geological and Lithological Structure and Hydrogeological Conditions

According to the data provided by the customer, the geological structure of the site where the research took place, up to the studied depth (10 m), includes deposits of the middle-upper Quaternary age (QII-III), represented by loams, covered from the surface by bulk soils (roadbed) and a ballast prism (Table 1). In hydrogeological terms, groundwater was detected, which was established during the survey period at a depth of 1.5 m from the base of the railway embankment. The executive layout of the geotechnical engineering borehole is shown in Fig. 3.

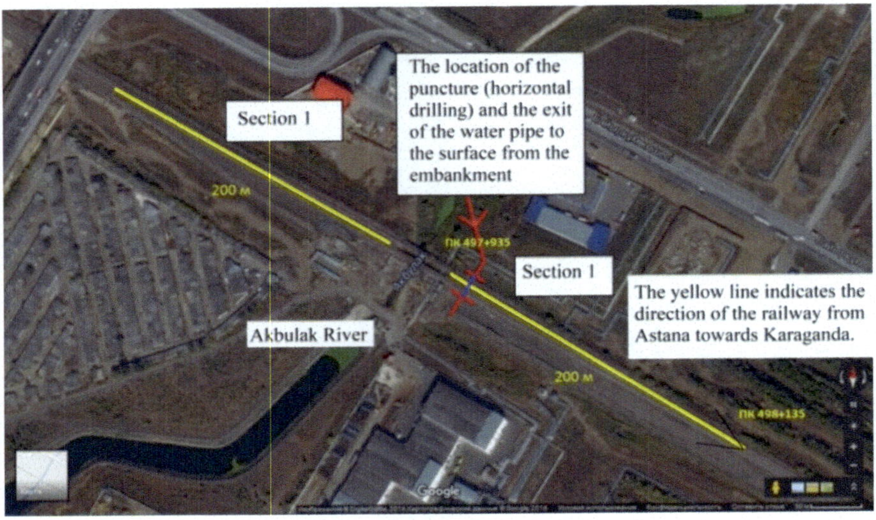

Fig. 2 Layout of the ground-penetrating radar survey (sections No. 1 and No. 2) of the railway embankment in Astana

Table 1 Engineering-geological characteristics of the designed construction site

No.	Layer thickness (m)	Geological and lithological structure	Groundwater level
1	0.0–0.37	Rubble (crushed stone)	The established groundwater level is 1.5 m in April
2	0.37–2.0	Subgrade (hard loam)	
3	2.0–3.6	Base: low-plastic loam with thin layers of sand	
4	3.6–5.0	Soft-plastic loam with layers of water-saturated sand	

The borehole was drilled by a self-propelled drilling rig PBU-2-319 on the basis of a Kamaz 5350-42 truck using the percussion-rock method with a diameter of 146 mm. During the drilling works, the appearance and recovery of groundwater level was monitored, and soil monoliths were taken [4–8]. Monoliths from the boreholes were sampled with a GK-123 soil carrier, 123 mm in diameter; soil samples with disturbed structure were taken in bags, and to preserve natural moisture content, they were taken in bouquets. Based on the results of drilling works, a geological and lithological description of the excavation was made (Table 1) and a geological and lithological column of the borehole was constructed (Fig. 3).

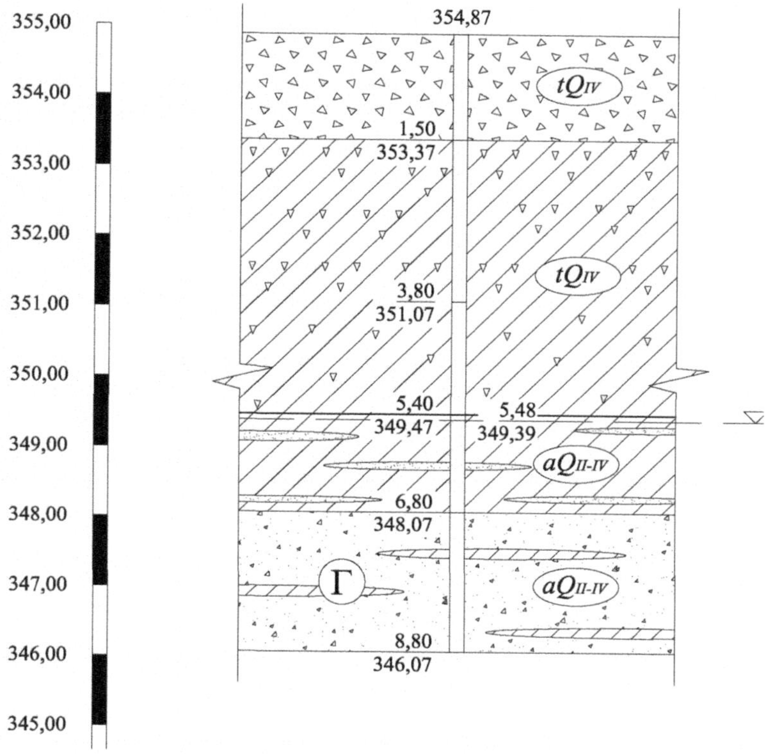

Fig. 3 Geologic and lithologic column of the well

3 Research Methods and Work Flow Diagram

The modern method of georadar probing was chosen as the most suitable radiophysical method of research both in depth and in resolution (Fig. 4). The georadar method is based on the study of the reflection of artificially created electromagnetic pulses from the boundaries of media with different electrical properties—permittivity and electrical conductivity. The frequency of the electromagnetic field used in georadars is usually in the range of 107–109 Hz. These frequencies are significantly higher than those used in classical geophysical electromagnetic methods, which allows for a much greater detail in the research, but limits the depth of the method. The attenuation of the electromagnetic field depends primarily on the frequency and electrical conductivity of the medium in which this field propagates. The lower the frequency and the lower the electrical conductivity, the deeper the electromagnetic pulses penetrate into the medium [4]. The task of studying a railway embankment is a classic one in the development of radar as a method of subsurface probing of the underlying environment. The technology of ground-penetrating radar on the surface usually involves the use of low-frequency antennas, which are located and moved directly along the surface using a handle rod (Fig. 4). The georadiolocation works were carried out to detect various deformations (mud bags, ballast troughs, etc.) using the radio-technical subsurface sounding device (GPR) "Oko-2," with AB-150 MHz and AB-400 MHz antenna unit manufactured by LLC "Logical Systems" (Moscow), along the profiles predetermined on the map. In total, 27 profiles were traversed along the entire length (300 m) of the railroad track (Fig. 5; Table 2).

Radar sounding by high-frequency block AB-150 MHz and AB-400 with the center frequency 400 MHz was carried out along the entire length of the railway embankment according to the technical specification (300 m). The use of the high-frequency block gives a gain in resolution with a loss in total sounding depth [10–15]. Figure 4 shows the radarogram of the profile after preprocessing. Analysis of the radarogram shows that the sounding depth is 9 m. The radarogram clearly traces the layer of ballast prism (up to 1 m), the layer of earth bed to the very base, represented by loams tight plastic loams. Also traced is the layer of the earth bed, represented by soft-plastic loams. The radarogram characterizes in detail the structural features of the upper part of the embankment [4]. Geophysical section by

Fig. 4 Ground investigation by geophysical method (GPR) [4]

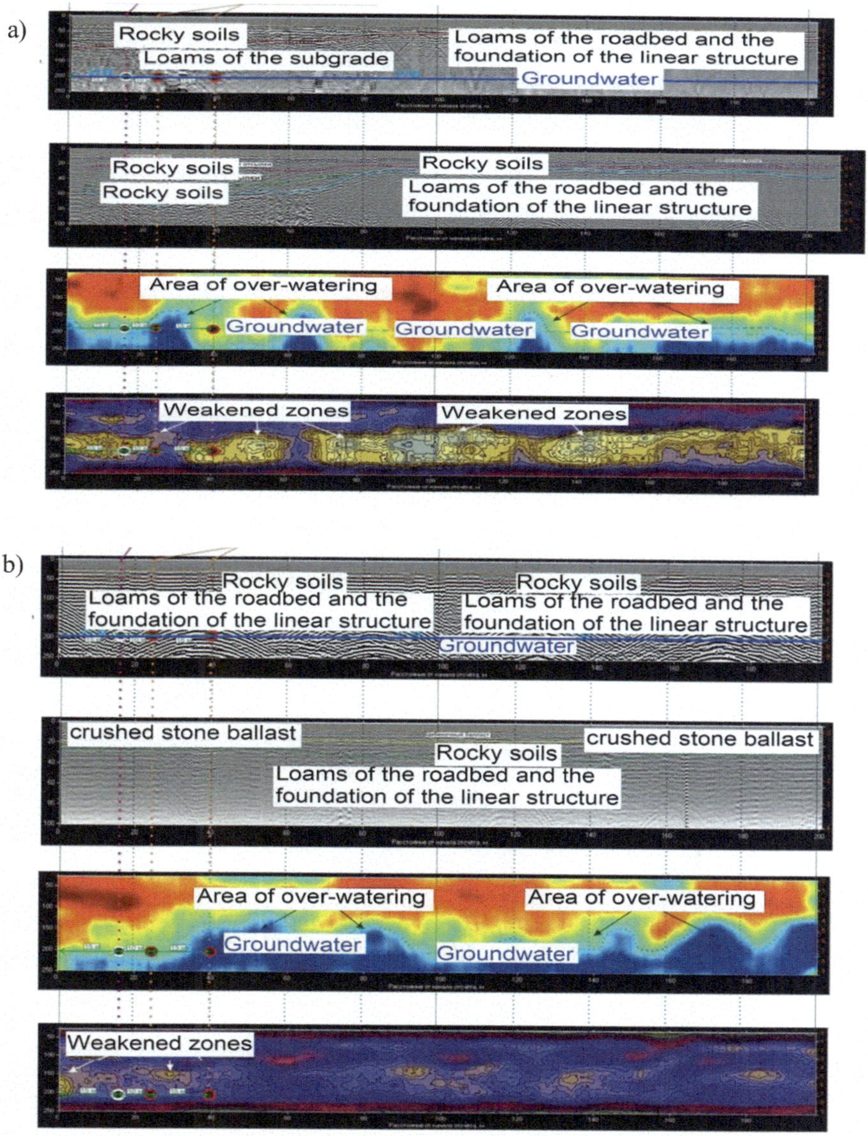

Fig. 5 Results of radarogram of area section 2: **a** section 2 left part of the route in the direction of Karaganda, **b** section 2 right part of the route in the direction of Karaganda, **c** section 2 middle of the railway line in the direction of Karaganda, **d** processing of results from 2-D to three-dimensional model [4]

c)

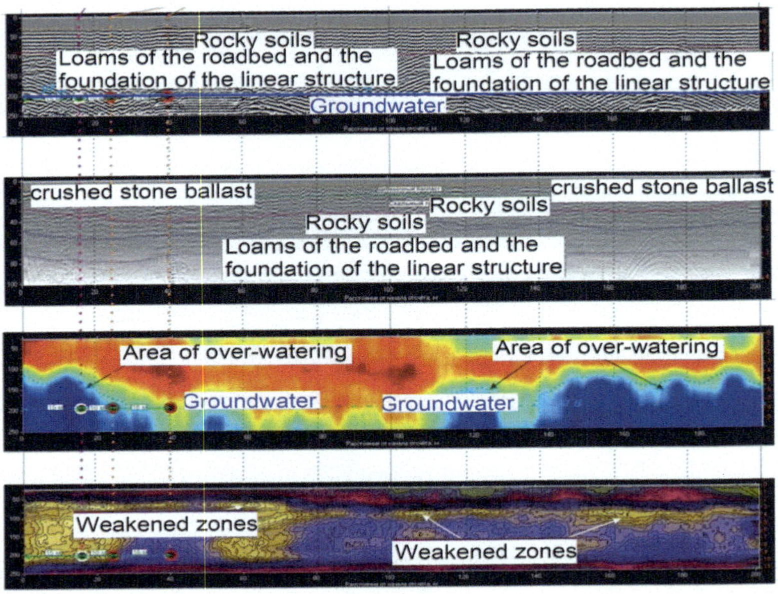

d)

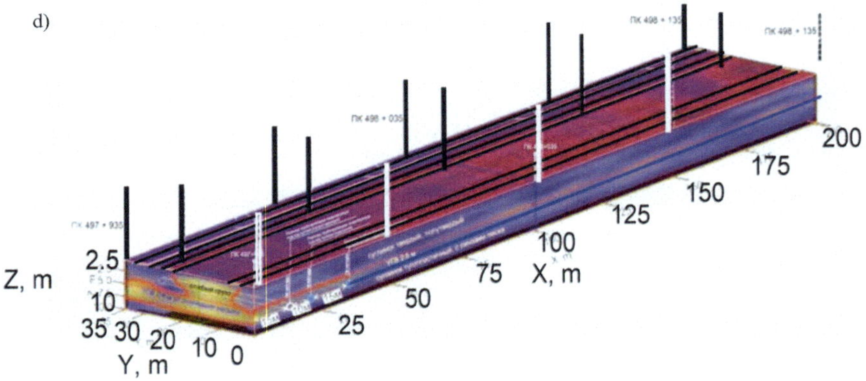

Fig. 5 (continued)

Table 2 Brief technical characteristics of antenna units of "Oko-2" [4]

Antenna unit type	Center frequency of transmission (MHz)	Maximum depth of geophysical survey (m)	Transmitting antenna excitation pulse amplitude
AB-150	150.0	12	Not less than 600 V
AB-400	400.0	5	Not less than 250 V

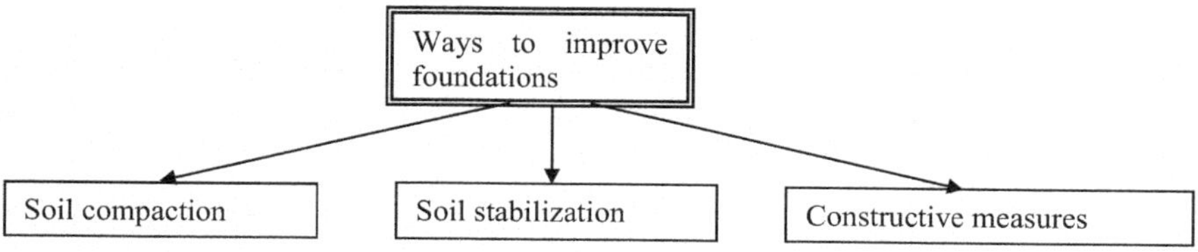

Fig. 6 Scheme of the main directions of development of ways to improve foundations

attribute "dielectric permittivity" characterizes the relative wetting of soils of the earth bed and embankment base (zones of light and dark blue colors).

4 Modern Methods of Reinforcement of Soil Foundations

The construction of linear engineering structures is inevitably associated with the development of new territories, including those located on weak soils. It often turns out that it is practically impossible to erect engineering structures on a natural foundation, due to the discrepancy between the characteristics of the foundation and the requirements. In these cases, either special design measures or methods of improving the foundation soils (artificial foundations) are used. There are three main directions for improving the foundation soils, shown in the diagram in Fig. 6.

5 Recommendation for the Use of Micropiles to Improve Soil Foundations

Using components with tools and technology intended for horizontal directional drilling is one method of fortifying the railway subgrade. This facility has the equipment necessary to place pipes beneath railroad tracks. Concrete-filled non-retrievable casing pipes with a diameter of 320 mm, for instance, might be employed. The walls of the pipes will range in thickness from 6 to 9 mm [4]. The geological and hydrogeological conditions at the site are taken into consideration when choosing the absolute markings for the axes used to lay the reinforcement elements. This is significant because, in order to prevent "leakage" of soil into the casing cavity, it is not advised to employ horizontal drilling equipment in loose, water-saturated soil. The subgrade may be destroyed as a result of such leaking [15–20].

It is suggested to lay casing in two horizontal rows with an axis variation of ± 50 tilt angles from the axis perpendicular to the railway track in order to maximize the efficiency of embankment reinforcement. This increases the total spatial rigidity of the reinforcing elements and eliminates the need for their arrangement in a single vertical plane [4].

Installing pipes from the bottom row in the order depicted in Fig. 7a–c is therefore advised; that is, pipes 1-1I, 3-3I, 5-5I, 2-2I, 4-4I, and pipes from the upper row, 6-6I, 8-8I, 10-10I, 7-7I, and 9-9I. Furthermore, prior to installing such pipelines, pilot boreholes should be drilled [4].

A steel tank of the appropriate volume filled with piece material (foundation blocks, prefabricated slabs, etc.) can be used as a thrust block to enable the rapid movement of drilling equipment to a different location while taking into account the exact marks of pipe axis resting above the subgrade base [21–25].

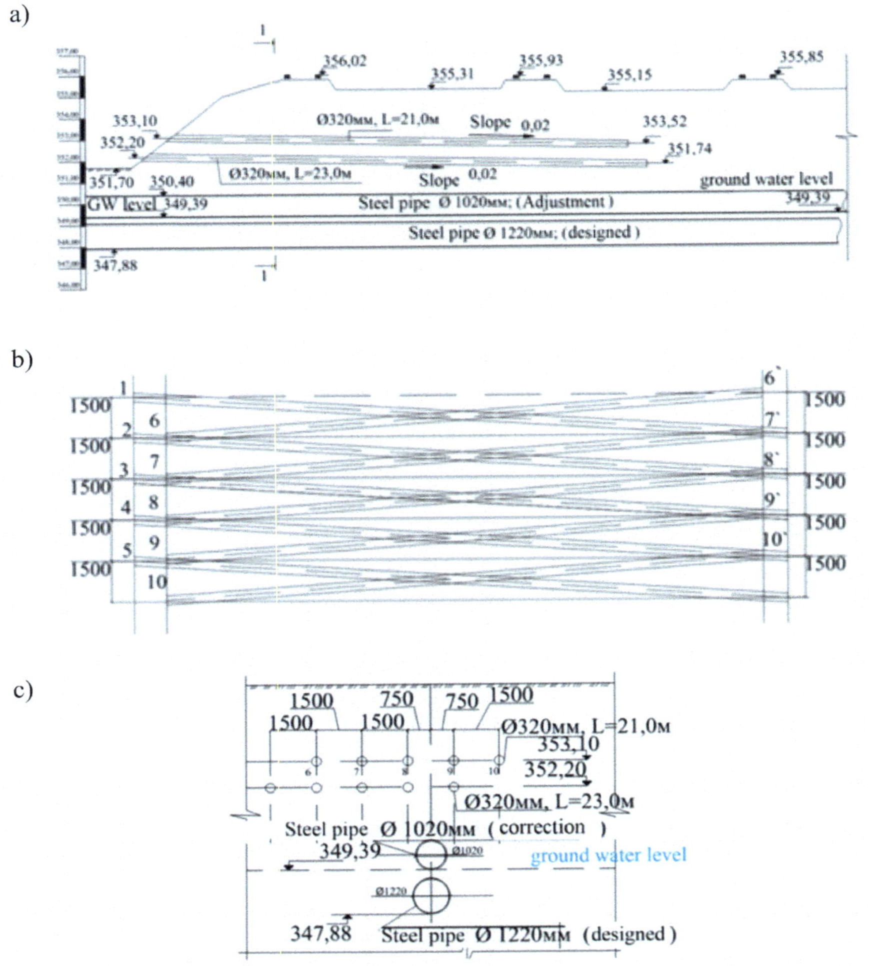

Fig. 7 At cross section 1-1, the railway subgrade is reinforced with horizontal concrete pipes: **a** vertical section of railway track; **b** horizontal section of railway track; **c** side drilling and installation of micropiles [4]

After installation in the design location, each pipe is filled to the full length with concrete of at least B25 class (25 MPa). Because the elastic modulus of heavy concrete class B25 is 1.4–1.6 times greater than that of, say, concrete class B10–B12.5, and 1.8–2.0 times higher than that of cement–sand solution M100, it is selected in order to increase the bending stiffness of pipe-concrete elements [4]. The tubular concrete elements' bending stiffness is increased by the increased modulus of elasticity, which reduces the subgrade's subsidence [26–31].

It is advised to keep an eye on the railway subgrade's subsidence process once the horizontal pipes are installed and, if required, to do any further repair and reinforcing work.

Modifications to the casing diameters and markings should be approved by the recommendations' developers.

6 The Second Method of Improving the Foundation Soils of the Railway Track

Installing inclined concrete pipes with rigid reinforcing is a second strategy for fortifying the railway subgrade. The idea behind this is to make sure that flowable soils below the groundwater table and solid and semisolid embankment loams have appropriate joints. When driven into the previously described soft soils, the inclined pipes are more robust, have greater bending stiffness, and are anchored in their upper section.

Given the characteristics of foundation elements with higher bending stiffness, like drilled piles (including inclined piles), their application in particular circumstances presents considerable challenges. Ensuring uninterrupted train movement is the primary concern when fortifying a railway subgrade. The height of the above electrical cables and their supporting components is a second issue.

It is advised to employ small-scale drilling rigs with technical features that enable the horizontal and inclined placement of reinforcing elements in water-saturated soils with the necessary geometrical specifications [4].

It is advised that the reinforcement pipes be arranged in the following order: Elements 1 through 10 should be put in the first stage such that they are perpendicular to the rail track axis, next in the order from items 11 to 22 as shown in Fig. 8 [4].

Following the immersion of the casing in the inner cavity, the 3–4 m long rigid reinforcement elements (rails) are lowered and joined to one another by welding steel plates: At the top, the dimensions are 45 mm in width, 150 mm in length, and 8 mm in thickness; at the bottom, they are 100 mm by 150 mm by 8 mm; and on the side edges in the vertical plane of the plate, they are 70 mm in height, 150 mm in length, and 8 mm in thickness on the two-sided plates' vertical edges of the permitted junction where connections require a minimum of four bolts [4].

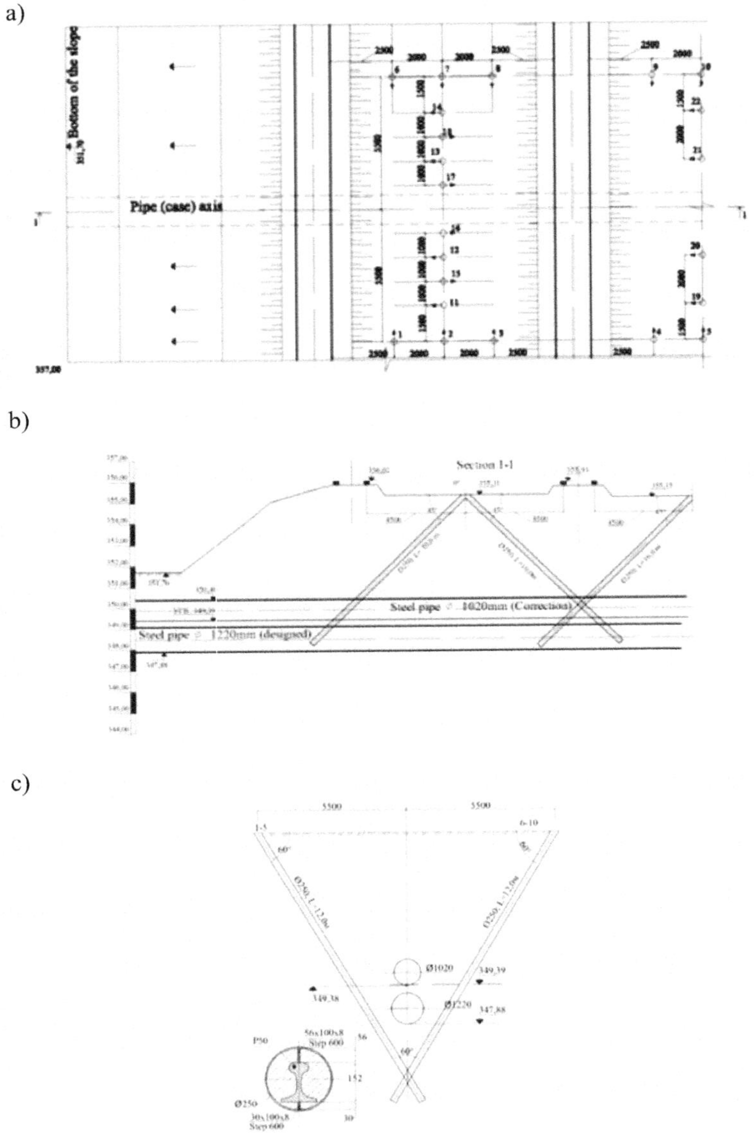

Fig. 8 At cross section 1-1, the railway subgrade is reinforced with vertical concrete pipes: **a** vertical section of railway track; **b** horizontal section of railway track; **c** side drilling and installation of micropiles [4]

With a pipe diameter of 250 mm initially in the longitudinal direction, the protective layer of steel plates was welded to the rails in vertical increments of 600 mm, with the lower zone measuring 30 mm in height, 100 mm in length, and 8 mm in thickness, and the upper face measuring 56 mm by 100 mm by 8 mm. It is advised to base the clamp height on the pipe's actual interior diameter (Fig. 8a–c).

It is permissible to make corrections to the casing's diameter (± 50 mm), the length of the reinforcement parts (± 1 m), and the slope angle (± 50 mm) without the developers' consent. Agreement of the guidelines' developers is required when modifying geometric parameters by values higher than those mentioned above [4].

7 Results and Discussions

1. Surface reinforcement is used only for compaction of soils with a low moisture content. It is performed by vibratory plates, rollers, rammers and, as a rule, is used in new construction or reinstallation of foundations.
2. Thermal consolidation (firing) is used mainly for stabilization of subsidence soils. Fuel is burned in hermetically sealed wells drilled in the thickness of the soil to be stabilized.
3. The most common physical and chemical types of soil reinforcement include cementation and the use of soil–cement materials. Soil cementation is based on the fact that cement or cement–sand mortar is pumped into the soil through injectors, ensuring the emergence of individual columns or massifs of cemented soil in the reinforced base. Cementation is usually used to strengthen sandy and coarse-grained soils.
4. Chemical methods of soil compaction and strengthening are silicification of all types, most often used in practice, ammonization, resinification, etc. The main material for silicification is liquid glass—a colloidal solution of sodium silicate. Ammonization consists of injecting gaseous ammonia into the soil under pressure. The method allows to impart non-subsidence properties to loess soils. Resinification is achieved by injecting aqueous solutions of synthetic resins into the soil.
5. It should be noted that the methods of strengthening soil foundations described above in a specific situation have significant limitations on the possibility of their application.

   The problems are as follows:

   - Technological schemes for the performance of works on strengthening soil foundations, the dimensions of machines and equipment when placing them on the surface of the roadbed are limited by the height of the contact wires, the presence of suspensions and supports of the contact network;

- Requirements for the continuity of technological processes and operations on soil strengthening on the one hand and the requirements for ensuring the continuous movement of trains along the railway line do not provide the opportunity to ensure high-quality performance of the work.

6. Analysis of the results of geotechnical studies and modern methods of strengthening soil foundations made it possible to determine the method of strengthening the railway embankment and develop recommendations for the restoration of a local section of the railway bed at the site of horizontal directional drilling work.

8 Conclusion

Experience in using ground-penetrating radar technology for probing a railway embankment in Astana has shown the following:

(1) The probing depth of the Oko-2 ground-penetrating radar with an AB-400 MHz antenna unit allows for the identification of reflective structures at a depth of about 9–10 m in the earth.
(2) A radar technique that allows for the effective study of lithological differences and deformation zones in various objects, including railway embankments.
(3) An area of increased moisture distribution in the soils of the railway embankment roadbed was identified in sections No. 1 and No. 2 from a depth of 7.0 m.
The spread of the waterlogged zone decreases from the right side of the embankment toward the left side.
    (a) On section No. 2 in the zone of transition of the water pipeline $D = 800$ mm under the railway tracks on PK 47 + 95.31 (496 km PK 9 + 54.05), the exit of groundwater to the surface of the foundation of the roadbed was detected based on the interpretation of geophysical radargrams (see graphic appendices, section No. 2). It is not recommended to cross the pipeline in the zone of weakened soils due to the spread of the groundwater level (GWL) over a significant distance on both sides of the Akbulak River.
(4) The groundwater level was detected (in sections No. 1 and No. 2) at a depth of about 7.1–7.4 m below the rail head.
(5) The section by the Q-factor attribute allowed us to identify weakened zones in the body of the roadbed and at the boundary of the base of the railway embankment (see graphic appendices No. 1 and No. 2 of the sections).
(6) In section No. 3, no anomalies were detected in the roadbed of the railway embankment. The roadbed is represented by clay soils, and the base is also composed of clay soils.

This article discussed the examination and possible corrective actions for a broken railroad track in Astana City, Kazakhstan, that is situated on soft soil. Two approaches to restore the base of the disturbed section of the railway bed were recommended, along with a description of methods and technologies, based on the results of engineering and geological surveys. These approaches are as follows: (1) install horizontal pipe-concrete elements to strengthen the railway embankment and (2) install inclined pipe-concrete elements with rigid reinforcement to strengthen the base of the railway embankment. Both strategies guarantee the safe execution of repairs on the damaged railway line in a high-voltage wire area without interfering with train operations.

Acknowledgements This research has been funded by the Committee of Science of the Ministry of Science and Higher Education of the Republic of Kazakhstan (Grant No. AP13268718).

References

1. Alves Costa P, Colaço A, Calçada R, Cardoso AS (2015) Critical speed of railway tracks. Detailed and simplified approaches. J Transp Geotech 2:30–46
2. Auersch L (2017) Mitigation of railway induced vibration at the track, in the transmission path through the soil and at the building. Procedia Eng 199:2312–2317
3. Dong K, Connolly DP, Laghrouche O, Woodward PK, Alves Costa P (2018) The stiffening of soft soils on railway lines. Transp Geotech 17:178–191. https://doi.org/10.1016/j.trgeo.2018.09.004
4. Zhussupbekov A, Omarov A, Zhumadilov I, Buranbayeva A, Drozdova I. Remediation proposals for damaged subgrade below railway line on soft soils. In: 20th international conference on soil mechanics and geotechnical engineering, Sydney, pp 5215–5218
5. Ferdous W, Manalo A (2014) Failures of mainline railway sleepers and suggested remedies—review of current practice. Eng Fail Anal 44:17–35
6. Kumaran G, Menon D, Nair KK (2003) Dynamic studies of railtrack sleepers in a track structure system. J Sound Vib 268(3):485–501
7. Powrie W, Yang LA, Clayton CRI (2007) Stress changes in the ground below ballasted railway track during train passage. Proc Inst Mech Eng J Rail Rapid Transit 221(2):247–262
8. Shih JY, Thompson DJ, Zervos A (2017) The influence of soil nonlinear properties on the track/ground vibration induced by trains running on soft ground. J Transp Geotech 11:1–16. https://doi.org/10.1016/j.trgeo.2017.03.001
9. Zhai W, He Z, Song X (2010) Prediction of high-speed train induced ground vibration based on train-track-ground system model. J Earthq Eng Eng Vib 9(4):545–554
10. Zhussupbekov A, Omarov A (2016) Geotechnical and construction considerations of pile foundations in problematical soils. In: Challenges and innovations in geotechnics—proceedings of the 8th Asian young geotechnical engineers conference, pp 27–32
11. Zhussupbekov A, Omarov A, Shakirova N, Razueva D (2020) Complex analysis of bored piles on LRT construction site in Astana. In: Lecture notes in civil engineering, vol 49, pp 461–471
12. Zhussupbekov A, Omarov A, Shakirova N, Abdrakhmanova B, Razueva D (2020) The experience of piling tests on Astana LRT construction site. In: 16th Asian regional conference on soil mechanics and geotechnical engineering, ARC 2019
13. Zhussupbekov A, Omarov A (2016) Modern advances in the field geotechnical testing investigations of pile foundations. Procedia Eng 16:88–95
14. Zhussupbekov AZ, Syrlybaev MK, Lukpanov RE, Omarov AR (2015) The applications of dynamic and static piling tests of Astana. In: 15th Asian regional conference on soil mechanics and geotechnical engineering, ARC 2015: new innovations and sustainability, pp 2726–2729
15. Zhussupbekov A, Omarov A, Zhukenova G, Tanyrbergenova G (2017) Pile foundations of mega projects: new railway station and LRT in problematical soil ground of Astana. Procedia Eng 189:511–518

16. Zhussupbekov AZ, Temirova FS, Riskulov AA, Omarov AR (2020) Investigations of historical cities of Uzbekistan and Kazakhstan as objects of the silk way. Int J Comput Civ Struct Eng 16(1):147–155
17. Zhussupbekov A, Mangushev R, Omarov A (2021) Geotechnical piling construction and testing on problematical soil ground of Kazakhstan and Russia. In: Lecture notes in civil engineering, vol 112, pp 89–107
18. Zhussupbekov A, Lukpanov R, Omarov A (2016) Bi-directional static load testing. Geotech Spec Publ GSP 265:35–42
19. Zhussupbekov A, Chang DW, Utepov Y, Borgekova K, Omarov A. Estimating the driven pile capacities for COF project in West Kazakhstan. Soil Mech Found Eng 52(2):121–127
20. Zhussupbekov A, Kaliakin V, Chang DW, Omarov A (2022) Investigation of interaction of piles at new cargo sea transportation route and LRT projects with problematic soils of Kazakhstan. In: Lecture notes in civil engineering, vol 164, pp 945–957
21. Zhussupbekov A, Mangushev R, Omarov A (2021) Geotechnical piling construction and testing on problematical soil ground of Kazakhstan and Russia. In: Lecture notes in civil engineering, vol 112, pp 89–107
22. Zhussupbekov A, Iwasaki Y, Omarov A, Tanyrbergenova G, Akhazhanov S (2019) Complex of static loading tests of bored piles. Int J GEOMATE 16(58):8–13
23. Zhussupbekov A, Omarov A, Morev I, Ashkey E, Borgekova K, Popov V (2017) Analysis results of static and dynamic loads tests of pile foundations in constructions site of expo-2017, Astana, Kazakhstan. In: ICSMGE 2017—19th international conference on soil mechanics and geotechnical engineering, Sept 2017, pp 3079–3082
24. Zhussupbekov A, Dhanya JS, Omarov A, Issakulov A, Iskakov S, Mukhanov D (2023) Model and field tests of drilled displacement system piles. In: 26th international scientific conference on construction the formation of living environment, FORM 2023, vol 410, p 03024
25. Issakulov A, Omarov A, Zhussupbekov A, Mussakhanova S, Issakulov B (2023) Investigation of the interaction of the bored micro pile by DDS (FDP) technology with the soil ground. Int J GEOMATE 24(105):11–17
26. Mussakhanova S, Zhussupbekov A, Omarov A, Abilmazhenov T, Issakulov A (2023) Features of testing piles for high-rise buildings in difficult soil conditions in Astana. Int J GEOMATE 25(110):106–113
27. Sultanov G, Zhusupbekov A, Lukpanov R, Enkebaev S (2010) Laboratory testing of elastic modulus of soil around of displacement pile. In: Proceedings of L.N. Gumilyov Eurasian National University, Astana, Kazakhstan, pp 214–219
28. Sultanov G, Zhusupbekov A, Lukpanov R, Enkebaev S (2010) Definition of dependence between experimental and designed values of bearing capacities of displacement pile. In: Proceedings of L.N. Gumilyov Eurasian National University, Astana, Kazakhstan, pp 226–232
29. Zhussupbekov A, Omarov A, Tanyrbergenova G (2019) Design of anchored diaphragm wall for deep excavation. Int J GEOMATE 16(58):139–144
30. Zhussupbekov A, Morev I, Omarov A, Borgekova K, Zhukenova G (2018) Geotechnical considerations of piling testing in problematical soils of West Kazakhstan. Int J GEOMATE 15(47):111–117
31. Omarov AR, Zhussupbekov AZ, Tulegulov AD, Zhukenova GA, Tanyrbergenova GK (2016) The analysis of the piling tests on construction site "the future of the free country". In: Challenges and innovations in geotechnics—proceedings of the 8th Asian young geotechnical engineers conference, 8AYGEC 2016, pp 127–130

Seismic Response of Tunnels in Soft Soil—Status and Future Prospects

Deepankar Choudhury and Rishav Baishya

Abstract For the safe functioning of tunnels, dynamic loading conditions, such as earthquakes, must be considered for design. The main idea here is to review tunnel responses under soft soil conditions and emphasize work done to date toward the seismic response of tunnels. Specifically, various parameters influencing tunneling, effects on tunnel structure, and ground failure due to seismic excitations are discussed. After that, a review of different closed-form solutions used for the transverse and longitudinal analysis of the response of tunnel structures is presented. Next is the study of various numerical models of tunnel systems with adjacent structures, both above and near the ground surface, like buildings, bridges, and buried pipelines. Here, the discussions are based on the different computational modeling procedures adopted by researchers and the effect of soil–tunnel structure interaction arising during seismic events. Subsequently, a review of interactions between underground tunnels and adjacent ground structures is also presented. The paper discusses the effects of segmental jointed tunnels under earthquake loads and the seismic vulnerability of tunnels. Finally, the article tries to stimulate ideas for new work in these areas for the researchers. Thus, emphasis is given to the identification of current gaps in understanding the seismic response of tunnels in soft soil.

Keywords Tunnels · Analytical methods · Numerical modeling · Seismic response · Soft soil

D. Choudhury (✉) · R. Baishya
Indian Institute of Technology Bombay, Mumbai, Maharashtra 400076, India
e-mail: dc@civil.iitb.ac.in

R. Baishya
e-mail: rishav11baishya@iitb.ac.in

B. Indraratna and C. Rujikiatkamjorn (eds.), *Recent Advances and Innovative Developments in Transportation Geotechnics*,
https://doi.org/10.1007/978-981-97-8245-1_10

1 Introduction

The construction of tunnels is a twenty-first century infrastructure demand for various utilities such as transportation, sanitation, irrigation, and storage facilities. The Tunnel Market Survey 2019 [1] estimates that globally, 5200 km of tunnels are being built annually, with more than three quarters of the demand arising from the Asian continent. Therefore, it is important to keep tunnels safe under any natural calamity like earthquakes, irrespective of the geological location of these structures.

Since the mid-seventies reports were being published about the damages on underground tunnels, however, there was no research on the study of interaction between tunnel and surrounding soil. It was only in 1993 that Wang [2] gave closed-form solutions for the determination of deformation, thrust, and bending moment of tunnel linings, thereby introducing the idea of relative flexibility for understanding the seismic responses of underground tunnels. However, damages to tunnel structures occurred multiple times during 1995 Kobe (Japan) earthquake, 1999 Chi-Chi (Taiwan) earthquake, 1999 Kocaeli and Duzce earthquake (Turkey), 2004 Niigata (Japan) earthquake, 2008 Wenchuan (China) earthquake, and 2010 Maule (Chile) earthquake [3–5]. Cases form these catastrophic events such as the failure of Daikai underground station during 1995 Japan earthquake [6], Bolu Tunnel in the Duzce earthquake [7], collapse of tunnel in Longxi Wenchuan event [8] and many others were later used as benchmarks for future research. These case studies play a vital role in validation of capabilities of present methods of analysis, material models, and interface conditions to mimic the actual seismic behaviors.

Laboratory tests such as centrifuge test and shaking table test are very useful in understanding observed responses of tunnels. Centrifuge tests done on scaled models are able to impose stress levels, similar to those which exist in the fields. However, selection of model material for the tunnels, minute detailing of the tunnel structures, and the excavation methods cannot be replicated exactly [9]. Shaking table test on the other hand has an added advantage of testing lager models with realistic materials, but the confining stresses used in these tests are much lower than actual field stresses, thus creating significant difference between the actual and recorded tunnel response [10]. To overcome the challenges posed by the above-mentioned laboratory tests, nowadays analytical and numerical methods are mostly used to accurately stand for actual conditions. As such the laboratory results are used for verification of analytical solutions and numerical models, with the latter having the same geometrical and boundary conditions as the tests.

The aim of the present article is to review the current developments for the understanding of the seismic behavior of tunnels. Section 1 gives an introduction on the requirements of incorporating seismic analysis for underground tunnel design; thereafter, Sect. 2 discusses about the different observed responses of tunnels under various seismic events recorded till date. It also discusses the mechanism involved behind observed responses for both longitudinal and transverse excitations for varying local site conditions, specifically for soft ground, along with discussions for earthquake-induced failure. Section 3 is about the various developments in seismic analysis of tunnels grouped into analytical, simplified, and dynamic analysis approach, along with discussions on future prospects. Discussion about fragility analysis of underground tunnels for its vulnerability evaluations is also presented. The final section concludes the review of different needs and approaches and shades light on the path for future research.

2 Seismic Response of Tunnels

2.1 Tunnels Observed Seismic Response

In the early twentieth century, tunnels were thought to be less prone to seismic damage. However, after several cases of tunnel failure, correlations were made between earthquake-induced failures and peak ground acceleration [11]. Similarly, correlations were also observed between tunnel damage and other parameters such as tunnel lining geometry, surrounding soil conditions, and earthquake characteristics. It was found that shallow tunnels were more likely to suffer damage during seismic events, with increasing earthquake magnitude and decreasing distance [12]. Using these correlations, the 1995 Kobe earthquake was marked as a major event with more than 10% of tunnel being close to the epicenter [13]. The 1999 Kocaeli earthquake which lead to collapsing of the twin Bolu tunnels was a failure in soft soil ground due to permeant ground deformation at shallow depths. Damage mostly occurred near slopes next to mined sections and portals due to seismic amplification. Most of the damage was caused to the under construction unfinished tunnel section that led to the ovaling of the transverse cross section due to accumulation of residual earth pressure on tunnel side walls [14]. Similar failures were observed in the other seismic events as mentioned in the previous section, for which detailed reviews are available in relevant publications [14]. From these tunnel failure reviews, it was found that the important parameters that affect the response of tunnels included the earthquake magnitude, hypo-central distance, underground soil conditions, tunnel burial depth, lining geometry, support systems, and sudden tunnel dimension change [15].

2.2 Mechanism of Seismic Response of Tunnels

Tunnel failures from past earthquakes have shown that tunnel structures should be designed to withstand strong ground shaking. Tunnel design should consider wave propagation effects, thus should be able to withstand seismically induced permanent ground deformations such as lateral spreading, liquefaction, landslides, and others. Axial deformations and longitudinal bending generally occur when earthquake shaking is aligned toward the longitudinal direction of the tunnel, similarly ovaling and combined racking-rocking deformations occur when shaking is along the transverse direction for circular and rectangular tunnels, respectively [16].

The seismic response of tunnels is different from those of aboveground structures as the kinematic force induced by surrounding soil is more prevalent as compared to inertial force form tunnel oscillations [2]. Thus, the response of tunnels is largely influenced by surrounding soil–structure interaction (SSI) during seismic excitations. The effect can be seen when assuming the behavior of soil and tunnel as elastic and considering a perfect bonding at the interface. Thus, rocking of tunnel cross section takes place around the section centroid, whereas considering nonlinearities at the interface and for the materials might result in movement of the rotation pole away from the centroid [17].

Responses of flexible tunnels in dry soil, such as post-shaking, residual earth pressure, lining shear stresses redistribution, due to soil yielding near interface, are captured if model nonlinearity is considered [18]. Soft ground when subjected to seismic excitations induces large deformations on flexible tunnels [19]. Sandy soil tends to amplify the earthquake shaking toward the surface, with dense sand showing higher amplification ratio for high frequency as compared to loosen sand at low frequency shakings [20]. Moreover, maximum principal stresses and strain occur at low frequencies with minimum value at crown and invert of tunnel [20]. Higher internal forces on segmented tunnel joints are visible at the tunnel invert when bottom part of tunnel is embedded in liquefiable soil layer for low frequency earthquakes [21]. However, racking deformation is more prominent in nonliquefiable layer, suggesting that shear deformation of interface soil induces larger deformations on tunnel side walls [21].

Analysis of subways station–tunnel connection for earthquake-induced liquefaction shows asymmetric settlement distribution along with accumulation of strain near the connection in the longitudinal direction [22]. When considering segmental tunnel linings with ring joints, deformation modes, such as joint opening, dislocation, and distortion, are observed [23]. Various methods are used for seismic evaluation of tunnels in the perpendicular direction, as well as for induced failures of ground, as the responses of tunnels are distinct in each of the cases. Many of these methods are being adopted by various design guidelines [24, 25].

3 Investigation of Tunnel Response in Seismic Conditions

3.1 Analytical Solutions

Analytical solutions are very useful for faster and easier design requirements. They generally serve the work of preliminary analysis as they are mostly based on various assumptions and simplifications. The earliest work on seismic design for underground structures was based on Newmark's closed-form solutions [26]. The design analysis used a free-field deformation approach for evaluating tunnel seismic response along with simple modifications to take care of soil–tunnel interactions [27]. Analytical solutions were also able to consider the full and no-slippage interface conditions [2], saturation in soil medium [28], and drained and undrained soil conditions [29]. Formulations of Wang [2] and Penzien [30] are mostly suitable for poor soil with low stiffness, which may include loose sand and undrained saturated clay [19]. Simplified ways were also developed for rectangular and circular tunnels by applying shear–strain distribution to the soil surrounding the underground tunnel [30] and actual deformations considering normal and shear components [31]. Even though most analytical methods are based on linear elasticity, nonlinearity, to some extent can be considered by using iterative processes, such as for strain-dependent modulus reduction for seismic analysis [32].

As analytical solutions are assumption based, therefore there are more accurate when the seismic deformations are small, mostly for low intensity excitations in stiff clays to rocks. Most researchers have compared the results of analytical solutions with results from numerical analysis to identify the possible gap areas and thus work upon them to overcome the challenges [33]. Moreover, as the transverse analysis is derived from plane strain assumptions, the results should be validated using numerical methods for better design outputs.

3.2 Simplified Solutions

For analysis along the transverse direction, various methods are available such as equivalent static procedure [25], simplified static analysis [2], frame-spring models [34], and pseudo-static coefficient deformation method [25]. For rectangular tunnels, the simplified static method considers tunnel deformations as functions of free-field ground displacements at tunnel level with a constant R (racking ratio) to account for SSI effects [2]. The calculated structural deformations are then applied on lining frame as equivalent static load or as a uniform pressure distribution. Flexibility of the tunnel linings with respect to surrounding soil were also incorporated in the racking ratio [30], and new relations were developed to mimic the coupled racking-rocking behavior of rectangular tunnels for transverse seismic excitations [35]. The frame-spring method used for transverse tunnel analysis considers the tunnel as beam element and the soil as spring [34]. Dynamic loading is introduced as (a) equivalent static inertia, (b) shear stress along tunnel perimeter, (c) earth pressure or ground displacement on tunnels walls. However, the above discussed methods do not show the effects of soil stress redistribution, due to which continuum model for soil is considered. Consider soil as continuum dynamic excitations which are modeled as equivalent static inertial force corresponding to free-field acceleration profile, which are determined using 1-D site response analysis. This technique is known as pseudo-static coefficient deformation method [25]. However, the equivalent force can also be introduced as ground displacements on model boundary or as displacement time histories, within pseudo-static frame of analysis.

For seismic analysis along the longitudinal direction, subgrade reaction method is used where tunnel is modeled as beam element on elastic soil springs, with spatial variation of earthquake excitations [25]. In this method when simplified equivalent static analysis is used, then the seismic excitation is applied as equivalent static displacement on springs, and when simplified dynamic analysis is done, then loading as displacement time histories is used. However, various challenges in this method are spacings of springs which is influenced by the wavelength and frequency of interest, and the magnitude of stiffness and damping for the selected springs and dashpots, respectively. Simplified solutions are essentially static, and thus, it has various challenges such as nonlinear soil response, varying soil element grid size, size boundary distance from tunnel cross section, and many more.

Various researchers have compared the results of simplified methods with those of experimental work and dynamic analysis and found that there might be either overestimation or underestimation of the results up to around 20–40% [36]. As there are no clear patterns of deviation from expected results, therefore more research is needed. Studies on effects of stress relief and redistribution during tunneling under seismic conditions have been carried out to some extent to shed light on predictability of simplified methods [37].

3.3 Dynamic Numerical Analysis Methods

In today's world, it is necessary to have accurate design models for real construction projects, where safety against any natural calamity is of utmost importance. It is also necessary to understand the exact reasons behind seismic failure of tunnels so that no such accidents occur in future. Thus, dynamic numerical analysis using earthquake time history is the most updated method for the analysis of soil–tunnel coupled model incorporating SSI effects [25]. Using various plasticity models via constitutive laws, nonlinearity in both the soil/tunnel material and interface contact can be modeled accurately. Numerical modeling is done for both design purposes and back calculation of observed tunnel behavior. These can be either based on (a) finite element/difference method, which is mostly used for modeling structures in continuum media and (b) discrete element method, which is for modeling discontinuous medium (fractured rock). Other types of methods include multi-scale models [38], coupled finite-boundary element models [39], and 2.5-D finite element–boundary element models [40]. Use of different soil constitutive models [41], tuning of soil damping [38], incorporating effect of nonlinear concrete lining [42], and nonsynchronous excitations for long tunnels [43] are some of the recent studies possible only through dynamic modeling. Full fluid–solid coupled analysis is performed to investigate the seismic response of tunnels in saturated sand for modeling undrained conditions and liquefaction phenomenon [21, 23].

In recent years, urban areas have faced a scarcity of land resources, which has led to a significant increase in the construction of high-rise buildings. Many countries are also actively planning and constructing comprehensive underground transportation networks, pipelines, air defense facilities, and more. The proximity of these structures to public services, and residential areas, as well as their interaction with aboveground building foundations, further complicates the situation. An integrated coupling system is formed by the connections between structures and aboveground constructions through the foundation soil. When an aboveground building is present, the surface acceleration response is increased, while the acceleration response is reduced when an underground structure is present directly above the station [44]. The explanation for the above distinction between the two can be attributed to the inertia impact of the aboveground construction and the isolation effect of the underground structure. The acceleration on ground surface is generally amplified compared to the input on bedrock, which is found to be reduced due to the presence of tunnels [45]. Nonetheless, the tunnel's presence has little effect on the site's natural frequency; in contrast, the natural frequency is significantly decreased for tunnel–soil–aboveground coupling systems [46].

3.4 *Fragility Analysis of Tunnels*

Fragility analysis of tunnels involves the evaluation of vulnerability of a tunnel to a specific damage state/intensity. Considering a region's seismicity (proneness to earthquakes) and the age and importance of tunnel systems in those regions, fragility analysis can help take informed decisions on pre-seismic structural renovations and post-earthquake managements. Functions that give the conditional probability of a structure meeting or exceeding various damage thresholds for various seismic intensities are used in fragility analysis. In the past, fragility curves were derived from damage states of earthquake hit tunnels [47], but recently, these curves have been possible through simplified/dynamic numerical analysis. For transverse loading, fragility curves have been developed, with seismic performance of tunnels mostly quantified by the ratio of earthquake-induced bending moment to the resisting moment [48]. Other quantifying parameters include permanent rotation between two lining segments [49], and displacement between crown and invert of tunnel [50]. For damage intensity, various measurement parameters may be used, such as free-field peak ground acceleration/velocity [51], arias intensity [52], and bedrock intensity [53]. Fragility analysis of tunnels in soft soil (low stiffness) conditions shows a higher probability of damage compared to stiffer ground conditions. From the above discussion, it can be found that there is limited work on seismic fragility analysis of tunnels and on earthquake-induced failure fragility curves.

4 Conclusions

Understanding the seismic behavior of subterranean tunnels has advanced significantly. Less research has been done, though, on how turbines react to failures brought on by earthquakes. Focus should also be toward considering asynchronous earthquake excitations. As most of the tunnel research considers the seismic excitation as S-waves, therefore there is a void in understanding the effect of surface waves (e.g., Rayleigh waves) on tunnels. The commonly followed approach is equivalent linear for tunnel lining concrete, which needs use of advanced constitutive modeling for detailed understanding of concrete response along with soil–tunnel nonlinear interface models. Interactions between multiple underground tunnels and with the aboveground structures also need to be focused on the possibility of multiple consecutive seismic excitations. Soil–tunnel models that take care of soil pore plasticity, heterogeneity, anisotropy, and nonlinear SSI need to be focused. Other areas of future work can also include tunnel lining age effect on seismic vulnerability of existing ones, as it can help in developing better retrofitting methods. Therefore, there exist vast scopes for understanding and improving seismic analysis of tunnels.

References

1. International Tunnelling and Underground Space Association—Association Internationale des Tunnels et de l'Espace Souterrain (2021) Tunnel market survey 2019, 3rd edn. Maison Internationale de l'Environnement
2. Wang JN (1993) Seismic design of tunnels: a state-of-the-art approach. Monograph 7. Parsons, Brinckerhoff, Quade and Douglas Inc.
3. Wang WL, Wang TT, Su JJ, Lin CH, Seng CR, Huang TH (2001) Assessment of damage in mountain tunnels due to the Taiwan Chi-Chi earthquake. Tunnel Undergr Space Technol 16(3):133–150
4. Elnashai AS, Gencturk B, Kwon OS, Al-Qadi IL, Hashash Y, Roesler JR, Kim SJ, Jeong SH, Dukes J, Valdivia A (2010) The Maule (Chile) earthquake of February 27, 2010, consequence assessment and case studies. Research report 10-04. Mid America Earthquake Center, Urbana, USA
5. Li T (2012) Damage to mountain tunnels related to the Wenchuan earthquake and some suggestions for aseismic tunnel construction. Bull Eng Geol Environ 71(2):297–308
6. Special issue of soils and foundations. Selected photographs on the 1995 Hyogoken-Nanbu earthquake. In: Special issue on geotechnical aspects of the January 17, 1995, Hyogoken-Nanbu earthquake. Soils Found (1996)
7. Kontoe S, Zdravkovic L, Potts D, Mentiki C (2008) Case study on seismic tunnel response. Can Geotech J 45:1743–1764
8. Yu HT, Chen JT, Yuan Y, Zhao X (2016) Seismic damage of mountain tunnels during the 5.12 Wenchuan earthquake. J Mt Sci 13(11):1958–1972
9. Hashash YM, Dashti S, Musgrove M, Gillis K, Walker M, Ellison K, Basarah YI (2018) Influence of tall buildings on seismic response of shallow underground structures. J Geotech Geoenviron Eng 144(12)
10. Abate G, Massimino MR, Maugeri M, Muir Wood D (2010) Numerical modeling of a shaking table test for soil-foundation-superstructure interaction by means of a soil constitutive model implemented in a FEM code. Geotech Geol Eng 28(1):37–59

11. Chen J, Luzhen J, Jie L, Xiaojun S (2012) Numerical simulation of shaking table test on utility tunnel under non-uniform earthquake excitation. Tunnel Undergr Space Technol 30:205–216
12. Dowding CH, Rozen A (1978) Damage to rock tunnels from earthquake shaking. J Geotech Eng Div ASCE 104:175–191
13. Asakura T, Sato Y (1996) Damage to mountain tunnels in hazard area. Soils Found 36(special issue):301–310
14. Lanzano G, Bilotta E, Russo G (2008) Tunnels under seismic loading: a review of damage case histories and protection method. In: Workshop of mitigation of the earthquake effects in towns and in industrial regional districts, Italy
15. Chen J, Luzhen J, Jie L, Xiaojun S (2012) Numerical simulation of shaking table test on utility tunnel under non-uniform earthquake excitation. Tunnel Undergr Space Technol 30:205–216
16. Owen GN, Scholl RE (1981) Earthquake engineering of large underground structures. Report no. FHWA/RD-80/195. Federal Highway Administration and National Science Foundation, 279 p
17. Tsinidis G (2017) Response characteristics of rectangular tunnels in soft soil subjected to transversal ground shaking. Tunnel Undergr Space Technol 62:1–22
18. Tsinidis G, Rovithis E, Pitilakis K, Chazelas JL (2016) Seismic response of box-type tunnels in soft soil: experimental and numerical investigation. Tunnel Undergr Space Technol 59:199–214
19. Zlatanović E, Lukić DČ, Prolović V, Bonić Z, Davidović N (2015) Comparative study on earthquake-induced soil–tunnel structure interaction effects under good and poor soil conditions. Eur J Environ Civ Eng 19(8):1000–1014
20. Asheghabadi MS, Matinmanesh H (2011) Finite element seismic analysis of cylindrical tunnel in sandy soils with consideration of soil-tunnel interaction. Procedia Eng 14:3162–3169
21. Shen Y, Zhong Z, Li L, Du X, El Naggar MH (2023) Seismic response of soil-shield tunnel systems in sandwiched liquefiable soil deposits. Tunnel Undergr Space Technol 140:105322
22. Haiyang Z, Xu W, Yu M, Erlei Y, Su C, Bin R, Guoxing C (2019) Seismic responses of a subway station and tunnel in a slightly inclined liquefiable ground through shaking table test. Soil Dyn Earthq Eng 116:371–385
23. Wu H, Ye Z, Zhang Y, Liu H, Liu H (2023) Seismic response of a shield tunnel crossing saturated sand deposits with different relative densities. Soil Dyn Earthq Eng 166:107790
24. Federal Highway Administration (1981) Earthquake engineering of large underground structures. Report no. FHWA/RD-80/195. Offices of Research & Development Structures and Applied Mechanics Division, Washington, DC
25. Federal Highway Administration (2009) Technical manual for design and construction of road tunnels—civil elements. Publication no. FHWA-NHI-10-034. Department of Transportation, Washington, DC
26. Newmark NM (1968) Problems in wave propagation in soil and rock. In: Proceedings of international symposium on wave propagation and dynamic properties of earth materials. University of New Mexico Press, New Mexico
27. St. John CM, Zahrah TF (1984) Seismic design considerations for underground structures. Adv Tunnel Technol Subsurf Use 4(3):105–112
28. Bobet A (2003) Effect of pore water pressure on tunnel support during static and seismic loading. Tunnel Undergr Space Technol 18(4):377–393
29. Bobet A (2010) Drained and undrained response of deep tunnels subjected to far-field shear loading. Tunnel Undergr Space Technol 25(1):21–31
30. Penzien J (2000) Seismically induced racking of tunnel linings. Earthq Eng Struct Dyn 29:683–691
31. Huo H, Bobet A, Fernández G, Ramírez J (2006) Analytical solution for deep rectangular structures subjected to far-field shear stresses. Tunnel Undergr Space Technol 21(6):613–625
32. Bobet A, Fernandez G, Huo H, Ramirez J (2008) A practical iterative procedure to estimate seismic-induced deformations of shallow rectangular structures. Can Geotech J 45:923–938
33. Kontoe S, Avgerinos V, Potts DM (2014) Numerical validation of analytical solutions and their use for equivalent-linear seismic analysis of circular tunnels. Soil Dyn Earthq Eng 66:206–219

34. Tsinidis G, Pitilakis K (2018) Improved R-F relations for the transversal seismic analysis of rectangular tunnels. Soil Dyn Earthq Eng 107:48–65
35. Tsinidis G, Pitilakis K, Anagnostopoulos C (2016) Circular tunnels in sand: dynamic response and efficiency of seismic analysis methods at extreme lining flexibilities. Bull Earthq Eng 14(10):2903–2929
36. International Organization for Standardization. ISO 23469 (2005) Bases for design of structures—seismic actions for designing geotechnical works. International Standard ISO TC98/SC3/WG10, Geneva, Switzerland
37. Gomes RC (2013) Effect of stress disturbance induced by construction on the seismic response of shallow bored tunnels. Comput Geotech 49:338–351
38. Sun QQ, Dias D (2018) Significance of Rayleigh damping in nonlinear numerical seismic analysis of tunnels. Soil Dyn Earthq Eng 115:489–494
39. Ding JH, Jin XL, Guo YZ, Li GG (2006) Numerical simulation for large-scale seismic response analysis of immersed tunnel. Eng Struct 28(10):1367–1377
40. Liang JW, Zhu J (2019) A FE-IBE method for linearized nonlinear soil-tunnel interaction in water saturated, poro-elastic half-space: I. Methodology and numerical examples. Soil Dyn Earthq Eng 120:454–467
41. Zhou S, He C, Guo P, Yu F (2019) Dynamic response of a segmented tunnel in saturated soil using a 2.5-D FE-BE methodology. Soil Dyn Earthq Eng 120:386–397
42. Kampas G, Knappett JA, Brown MJ, Anastasopoulos I, Nikitas N, Fuentes R (2019) The effect of tunnel lining modelling approaches on the seismic response of sprayed concrete tunnels in coarse-grained soils. Soil Dyn Earthq Eng 117:122–137
43. Fabozzi S, Bilotta E, Yu H, Yuan Y (2019) Effects non-uniform seismic shaking on soil-tunnel interaction. In: 7th international conference on earthquake geotechnical engineering. AGI, Rome, Italy
44. Wang GB, Wu HS, Wang YX (2016) Study on seismic responses of subway stations and interaction system with upper steel frame structure. J Build Struct 37:79–84
45. Abate G, Corsico S, Massimino MR (2016) FEM modelling of the seismic behavior of a tunnel-soil-aboveground building system: a case history in Catania (Italy). Procedia Eng 158:380–385
46. Wang GB, Yu YL, He W (2014) Seismic response of interaction system of underlying tunnels, soils, and adjacent frame structures. Chin J Geotech Eng 36(2):334–338
47. Corigliano M, Lai C, Barla G (2007) Seismic vulnerability of rock tunnels using fragility curves. In: 11th international society for rock mechanics congress. Lisbon, Portugal
48. Argyroudis S, Pitilakis K (2012) Seismic fragility curves of shallow tunnels in alluvial deposits. Soil Dyn Earthq Eng 35:1–12
49. Fabozzi S, Bilotta E, Lanzano G (2017) A numerical study on seismic vulnerability of tunnel linings. In: Proceedings of 3rd performance based design in earthquake geotechnical engineering, Vancouver, Canada, 16–19 July 2017
50. Andreotti G, Lai C (2019) Use of fragility curves to assess the seismic vulnerability in the risk analysis of mountain tunnels. Tunnel Undergr Space Technol 91
51. Nguyen DD, Park D, Shamsher S, Nguyen VQ, Lee TH (2019) Seismic vulnerability assessment of rectangular cut-and-cover subway tunnels. Tunnel Undergr Space Technol 86:247–261
52. Huang G, Qiu W, Zhang J (2017) Modelling seismic fragility of a rock mountain tunnel based on support vector machine. Soil Dyn Earthq Eng 102:160–171
53. de Silva F, Fabozzi S, Nikitas N, Bilotta E, Fuentes R (2021) Seismic vulnerability of circular tunnels in sand. Géotechnique 71(11):1056–1070

A 10-Year Research on Sustainable Pavement Materials

Suksun Horpibulsuk, Menglim Hoy, Apichat Suddepong, Artit Udomchai, Apinun Buritatum, Teerasak Yaowarat, Kongsak Akkharawongwhatthana, Chakkrid Yeanyong, Jitwadee Horpibulsuk, Manilika Mobkrathok, Punvalai Choenklang, Apisit Laomuad, Karn Karntatam, Avirut Chinkulkijniwat, and Arul Arulrajah

Abstract This paper summarizes 10-year research on sustainable road construction by using recycled materials such as construction and demolition (C&D) materials; waste by-products, which are fly ash, steel slag, and calcium carbide residue; synthetic and natural fiber-reinforcement; and synthetic and natural polymers in various pavement applications. Advanced techniques were used to analyze the mechanical properties and suitability of these materials as replacements for virgin aggregates. The research identifies optimal material compositions, demonstrating improved pavement performance and environmental and economic benefits. Findings highlight the potential of using recycled materials like bottom ash and natural fibers, advocating for sustainable and cost-effective road construction practices. The research contributes to the development of durable and environmentally friendly road infrastructure, with implications for future urban and high-traffic applications.

Keywords Sustainable pavement · Recycled materials · Pavement performance

1 Introduction

Roads significantly impact the environment through intensive resource and aggregate use. With projected global road expansion, increased dependency on virgin extraction and non-renewables further drives ecological damage. Urgent transition toward

S. Horpibulsuk (✉) · M. Hoy · A. Suddepong · A. Udomchai · A. Buritatum · T. Yaowarat · K. Akkharawongwhatthana · C. Yeanyong · J. Horpibulsuk · M. Mobkrathok · P. Choenklang · A. Laomuad · K. Karntatam · A. Chinkulkijniwat
Suranaree University of Technology, Nakhon Ratchasima 30000, Thailand
e-mail: suksun@g.sut.ac.th

S. Horpibulsuk
The Royal Society of Thailand, Bangkok 10300, Thailand

A. Arulrajah
Swinburne University of Technology, Melbourne, Australia

B. Indraratna and C. Rujikiatkamjorn (eds.), *Recent Advances and Innovative Developments in Transportation Geotechnics*,
https://doi.org/10.1007/978-981-97-8245-1_11

demonstrably promising and rigorously sustainable recycled construction materials is imperative for next-generation green road infrastructure. Traditional road-building methods heavily depend on the extraction and processing of virgin aggregates, primarily crushed rock, sand, and gravel. This process disrupts natural ecosystems, contributes to air and water pollution, and generates substantial greenhouse gas emissions. Moreover, the extraction of these resources is often energy-intensive and involves considerable transportation costs.

To combat these environmental concerns, the construction industry has turned its attention to utilizing recycled materials as alternatives to virgin aggregates. The construction and demolition (C&D) sector generates a significant amount of waste materials, such as recycled concrete aggregate (RCA), recycled asphalt pavement (RAP), waste brick, glass, and plastics, that can be processed and reused in road construction. Industrial by-products, such as fly ash and slag, also offer promising alternatives to virgin materials.

Numerous feasibility and performance studies have demonstrated the successful integration of recycled materials in various pavement applications. These recycled materials have been shown to have promising performance characteristics when used as partial or full replacement for virgin aggregates.

This paper presents the systematic research outcomes of a 10-year study investigating the performance of utilization of various recycled waste materials in rigid and flexural pavement applications. The central aim of this research is that utilizing locally available recycled materials as alternative pavement construction materials can significantly enhance the sustainability and resource efficiency of road-building practices. By reducing the reliance on virgin aggregates, these recycled materials not only conserve natural resources but also mitigate the environmental impacts associated with their extraction, processing, and transportation. The findings of this study will contribute to the advancement of sustainable road construction practices and provide valuable insights for policymakers and engineers seeking to minimize the environmental footprint of road infrastructure development.

2 Materials and Methods

2.1 Materials

A range of recycled materials was investigated as substitutes for conventional aggregates in pavement construction. These included C&D wastes, RAP, RCA, crushed clay brick (CB), granulated blast furnace slag (GBFS), waste rock (WR), fine recycled glass particles (FRG), medium recycled glass particles (MRG), Polyethylene terephthalate (PET) waste, and crumb rubber (CR). Waste from industrial by-products: fly ash (FA), granulated blast furnace slag (S).

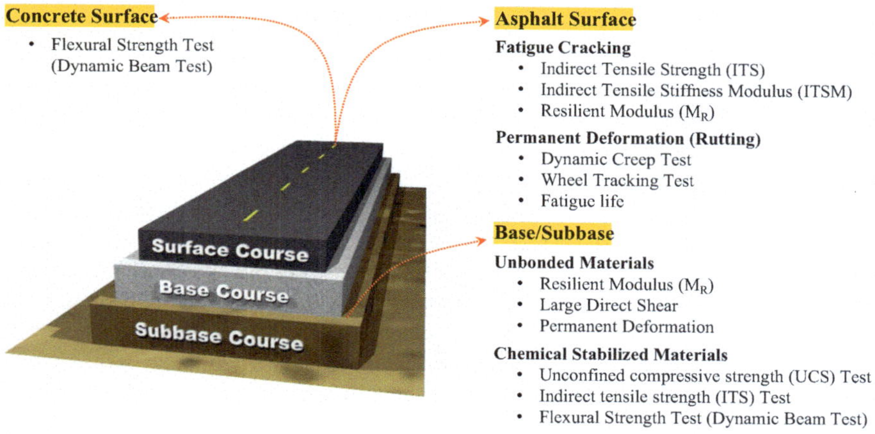

Fig. 1 Schematic illustrates the performance tests for recycled waste materials in each layer of pavement

2.2 *Performance Methods*

The methodological approach of our decade-long research into sustainable road engineering has been multifaceted, incorporating rigorous laboratory experiments, analytical modeling, and field performance monitoring to evaluate the sustainability of recycled materials for use in road construction. Figure 1 summarizes the experimental framework assessing recycled material blends in each pavement layer. Advanced analytical techniques, including X-ray fluorescence (XRF), X-ray diffraction analysis (XRD), and scanning electron microscopy (SEM) were employed to ensure a thorough understanding of the fundamental properties of the recycled materials used in pavement layers.

3 Results and Discussion

3.1 *Unbound Base/Subbase Materials*

The use of fine recycled glass (FRG), recycled concrete aggregate (RCA), and waste rock (WR) in an unbound pavement base/subbase was investigated. The study examined blends of FRG/RCA and FRG/WR for base applications using lab tests, including repeated load triaxial tests and field tests. The optimal blend contained 20% FRG, offering workability and strength. Field tests suggest FRG blends are suitable for subbase layers, but may not meet all specifications for base materials [1]. The physical and shear strength properties of various C&D materials, including RCA, CB, RAP, WR, FRG, and MRG for pavement applications were evaluated.

The shear strength behaviors fell into two categories: those with dilatancy-induced peak strength (RCA, WR, CB) and those exhibiting strain-hardening (RAP, FRG, MRG). The study concluded that compacted C&D materials, especially RCA, CB, and WR, are suitable for pavement base/subbase applications, meeting necessary friction angle criteria [2].

The influence of traditional stabilizers (cement and bitumen) and nontraditional stabilizers (organic petroleum and synthetic polymer) on pavement base materials was investigated through cyclic triaxial tests and freeze–thaw cycles. The freeze–thaw actions exert almost no effect on the samples for both types of stabilizers. The studied stabilizers can significantly reduce the thickness of a traditional untreated base layer by an average value of 50%. Stabilizer choice greatly impacts the cost of treated base courses. While innovative stabilizers demonstrate superior mechanical performance, their high prices limit cost-competitiveness versus traditional base options. However, most stabilized technologies remain economical in comparison with UGM bases, except when aggregate transportation distances are great, increasing costs [3].

Sustainable usage of C&D materials in pavement and railway structures was evaluated via experimental analysis and discrete element method (DEM) simulation. Results show particle shape significantly influences crushing strength, and the DEM simulations align well with lab experiments. Although the flaky assembly had the lowest initial porosity and densest packing structure, this sample experienced the worst fragmentation of agglomerates and the highest permanent strain under the one-dimensional compression test [4].

The impact of incorporating recovered plastics (RP) into C&D material blends was explored, focusing on their compressibility and stress-dilatancy responses. Compressibility and potential for collapse were assessed through one-dimensional consolidation tests, while shear strength and stress-dilatancy responses were evaluated using monotonic drained triaxial tests. Findings indicate that RP content primarily affects compressibility, with a lesser impact on C&D aggregates' collapse potential. However, increasing RP content reduces the stiffness and strength of the blends. RP inclusion improves energy absorption, as evidenced by increased total work input during shearing [5].

3.2 Cement-Stabilized Base/Subbase Materials

The use of RG, CB, and RCA in triple blends for pavement construction was investigated. The study aimed to maximize the use of these materials in cement-stabilized base and subbase applications. Seven blends with RG content varying from 0 to 40%, combined with 20% CB and up to 80% RCA, were assessed with 3% cement. The blend meets local road authority specifications for unconfined compressive strength, modulus of rupture, and flexural beam fatigue. Blends with up to 30% RG satisfy requirements for pavement layers with a modulus under 500 MPa. Wheel-tracking tests indicate that the 15% RG blend has excellent rutting resistance, suggesting a potential to increase RG content. The study supports using up to 35% combined RG and crushed brick (CB) with recycled concrete aggregate in pavements designed for a modulus between 500 and 3500 MPa, aligning with sustainable construction goals [6].

Cement-treated CR and RG blends for pavement construction were evaluated. Results indicated that UCS exceeded 3.5 MPa after 7 days, with notable strength increases at 28 days, especially for CR70/RG30 blend. Repeated load triaxial tests indicated CR100 had the highest resilient modulus; CR70/RG30 had the lowest. Modulus of rupture ranged from 0.78 to 0.94 MPa, with flexural modulus between 2465 and 3503 MPa. Up to 15% RG is initially recommended for use in cement-treated CR pavement subbase applications [7].

The enhancement of RCA for road construction using polyvinyl alcohol (PVA) and cement was investigated. PVA addition lowered M_R, but values remained above the minimum standard. The California Bearing Ratio (CBR) and swelling in soaked conditions decreased with increased PVA content. A 3% cement-stabilized RCA with up to 1.5% or 2% PVA showed physical and strength properties meeting road authority requirements. However, it was found that adding more than 1.5% PVA to cement-stabilized RCA significantly improves UCS and modulus of toughness [8].

The utilization of garnet residues (GR) and tire-derived aggregates (TDA) to enhance soft clay (SC) for subgrade construction was examined. It indicated that the optimal mix: SC:GR = 90:10, with 2% cement and 2% TDA, offers the best performance. Strength increases with cement up to 2%, then plateaus due to hydration limits. The findings support GR and TDA use for a sustainable, high-fatigue-resistant subgrade material [9].

3.3 *Geopolymer-Stabilized Base/Subbase Materials*

The strength, wet-dry durability, and toxic leaching investigations were performed to examine the utilization of RAP–fly ash-based geopolymers as a sustainable pavement base material. The 28-day UCS values of RAP-FA geopolymers are significantly higher than the RAP-FA blend, attributed to geopolymerization reactions. N-A-S-H gel and C-A-S-H products coexist, leading to increased UCS values with a decreasing $NaOH/Na_2SiO_3$ ratio. Strength increases with up to 6 wet-dry cycles due to chemical reactions forming C-S-H and C-A-H, then decreases due to moisture loss. The leachability of heavy metals from these materials was measured and found no harm when compared with international standards [10–12].

The use of FA and S-based geopolymer-stabilized washed recycled sand (WRS) as a sustainable alternative to virgin sand in pavement subgrade materials was examined. Samples with S generally exhibited higher UCS than those with FA. However, S-based geopolymers required significant water, leading to a strength drop in samples with excess S at high curing temperatures, or extended curing periods. The best UCS (22 MPa) was achieved with a mix of 15% FA + 15% S, cured for 7 days at 40 °C. The suitability of geopolymer-stabilized RS as a substitute for quarried sands in the pavement was found when stabilized with 30% FA or 10% S and cured for

7 days at 21 °C [13]. FA geopolymer-stabilized CR using asphalt emulsion (AE) for sustainable pavement base material was examined. UCS of FA geopolymer-stabilized CR increased with increased FA content and curing time. The addition of AE was found to retard the geopolymerization reaction, reducing UCS. However, FA geopolymer-stabilized CR with AE exhibited higher FS, ITS, ITRM, and ITFL, along with a lower carbon footprint [14].

Stabilization of PET plastic and C&D waste blends using FA and S-based geopolymers in light traffic road bases/subbases was evaluated. The UCS of all geopolymer-stabilized blends using 5% FA + 5% S and 10% S met the minimum requirements for light traffic road bases/subbases. From the 12 geopolymer-stabilized blends evaluated for their M_R values by RLT tests, 10 blends compliant with the minimum MR limit of 80 MPa for the pavement bases and the other two blends, 5% PET + 95% RCA + 5% FA + 5% S and 5% PET + 95% CB + 10% FA satisfied the minimum M_R requirement of 35 MPa for the pavement subbases [15].

3.4 Cement-NRL Stabilized Base/Subbase Materials

The use of natural rubber latex (NRL) as an additive in cement stabilization for unpaved roads was attested, which focuses on its impact on strength, fatigue life, and environmental benefits. Increasing NRL replacement ratios enhanced UCS, ITS, and NF, with maximum values at optimal NRL ratios. NRL addition increased total deformation under fatigue stress but significantly reduced plastic deformation, indicating enhanced resistance to fatigue-induced plastic strain. Cement-NRL-stabilized soil showed lower CO_2 emissions compared to traditional cement-stabilized soil at similar ITS levels. NRL replacement could reduce the thickness of cement-NRL-stabilized base courses for given traffic volumes and service life, potentially lowering construction costs by 17.26% compared to cement-stabilized bases [16, 17].

The investigation of the influence of NRL on the performance of cement-stabilized RCA as a pavement base material was also performed. An optimal 10% *r*/*c* ratio in 3% cement-stabilized RCA (3C10R-RCA) showed the best performance, with reduced permanent deformation, fatigue cracking, and rutting susceptibility. The fatigue life of 3C10R-RCA was 79.3% higher than standard 3% cement-stabilized RCA (3C-RCA), and rut depth was 8% lower. At the optimal 10% *r*/*c* ratio, cement-NRL stabilized RCA maintained higher UCS values even after 12 w-d cycles, surpassing the 2.413 MPa requirement for high-volume traffic roads [18, 19]. In addition, the effect of NRL on improving the properties of cement-stabilized lateritic soil and recycled materials, steel slag (SS), and RCA, for sustainable pavement base applications was also examined. NRL additive significantly improved the fatigue life and rutting resistance of cement-stabilized materials, with the formation of NRL films acting as a crack-bridging mechanism [20, 21].

3.5 *Geogrid Reinforced C&D Waste for Pavement Base Application*

The use of natural kenaf geogrids for reinforcing RAP and RCA in sustainable pavement construction was investigated, focusing on their interface shear strength. The research analyzed how the gradation of RAP and RCA particles and the aperture sizes of the geogrid (*D*) affect the interface shear strength under varying normal stresses. A key parameter, the *D*/FD ratio (FD being particles finer than the geogrid aperture), was identified as critical in governing interface shear strength. A generalized equation, $\alpha = a(D/\mathrm{FD}) + b$ (where α is the interface shear strength coefficient), was developed for predicting interface shear strength. For kenaf geogrid-reinforced RCA, a higher *D* led to an increased interface friction angle (δ) and adhesion, with the effect varying based on the size of the RCA particles [22].

3.6 *Foam Bitumen Stabilized Base/Subbase Materials*

This section focuses on the application of foamed bitumen (FB) stabilization for C&D waste blends in pavement base layers, assessing their strength and deformation responses. ITS, UCS, and CBR values generally decreased with an increase in RAP, RG, and RP content. However, FB stabilization notably reduced the permanent deformation, particularly in blends with higher percentages of these materials. The permanent deformation responses were analyzed using the shakedown concept, indicating that FB-stabilized C&D wastes have stable responses under cyclic loads [23].

3.7 *Sustainable Concrete Pavement Applications*

NRL was used as a green additive to improve the flexural strength properties of concrete pavements under cyclic traffic loading. The results indicated that the UCS of NRL-concrete decreased with increasing the r/c ratio at all w/c ratios ($w/c = 0.3$, 0.4, and 0.5) and curing times, while the highest flexural strengths were found at the optimal $r/c = 0.58\%$, 1.16%, and 1.73% for $w/c = 0.3$, 0.4, and 0.5, respectively. The $w/c \leq 0.4$ and $r/c \leq 2.31$; $w/c = 0.5$ and $r/c = 0.58$–1.73 were found to be applicable to NRL-concrete pavements, as the compressive and flexural strengths met the minimum requirements specified by the Department of Highways (DOHs), Thailand. Concrete modified by NRL at the $r/c \leq$ optimum ratio had lower plastic deformation than normal concrete at the same applied flexural stress, hence, higher fatigue life [24].

This research explores the use of PET and crumb rubber as aggregate replacements in NRL-modified concrete, aiming to leverage the benefits of these materials for sustainable construction. An increase in the replacement ratios of PET and crumb rubber led to a decrease in the compressive and flexural strengths of the NRL-modified concretes. The integration of PET and crumb rubber aggregates in the concrete mix enhanced its ductility and energy absorption capabilities. The addition of NRL was found to improve bond strength between aggregates and hydration products, acting as a bridge in NRL-modified concrete. Up to 10% replacement of aggregates with PET and crumb rubber was deemed suitable for NRL-modified concrete in rigid pavement applications, meeting the mechanical strength requirements of the DOHs, Thailand [25].

The influence of polyvinyl alcohol (PVA) on the flexural strength of concrete containing RCA as the rigid pavement was conducted. The research investigated how varying PVA-to-cement (*p*/*c*) and water-to-cement (*w*/*c*) ratios affect the compressive and flexural strengths of RCA-PVA concrete. The study observed that flexural strength increased with *p*/*c* ratios up to an optimum value, beyond which it decreased. The research concluded that *p*/*c* ratios ≤ 1 at *w*/*c* ratios ≤ 0.5 are suitable for developing RCA-PVA concrete for rigid pavements. In addition, the study investigates the impact of PVA to binder (*p*/*b*) and FA to cement (FA/*c*) ratios on the mechanical strength of RCA-PVA-FA concrete was also presented. Adding PVA reduced compressive strength but significantly increased flexural and tensile strengths at an optimal *p*/*b* ratio of 1%. PVA films improved the reinforcing ability within the cementitious matrixes, but excess PVA hindered hydration and reduced tensile bonding [26].

3.8 Sustainable Asphalt Pavement Applications

This research evaluates the performance, field behavior, and cost-effectiveness of fiber-reinforced asphalt concrete (FR-AC) pavements using both conventional AC60/70 and polymer-modified asphalt (PMA) binders. The study included indirect tensile resilient modulus, indirect tensile strength modulus, fatigue life, dynamic creep, and wheel-tracker tests. Field trials for AC60/70 and PMA mixtures, both with and without fibers, assessed the International Roughness Index (IRI), texture depth, and rutting over time. Fiber addition improved fatigue cracking and rutting in both AC60/70 and PMA mixtures. The improvement was more pronounced for fatigue cracking in PMA mixtures. The PMA + Fiber mixture exhibited the best overall performance in laboratory and field tests compared to other materials. The AC60/70 + Fiber mixture was the most economical, with cost increases of 10%, 53%, and 68% for AC60/70 + Fiber, PMA, and PMA + Fiber, respectively [27].

This research delves into how aggregate properties and NRL additives impact the performance of natural rubber-modified asphalt (NRMA) concrete mixtures. NRMA mixtures were prepared using three types of aggregates (granite, limestone, and basalt), asphalt cement AC60/70, and varying rubber-to-binder (*R*/*B*) ratios (3, 5, and

7%). A 3% *R*/*B* ratio was identified as optimal, offering the best balance of mechanical properties and performance in NRMA mixtures. Basalt and granite, with their high stability, contributed to increased resilient modulus, rutting resistance, and reduced permanent deformation in NRMA mixtures. Granite, due to its superior strength and shape parameters, significantly enhanced the fatigue life of NRMA mixtures compared to other aggregates. The primary mineral compounds in limestone (CaO, MgO, and others) affected the indirect tensile strength and skid resistance of NRMA mixtures [28].

Porous asphalt concrete (PAC), an open-graded asphalt concrete with high air void content, offering permeability and high surface frictional resistance was invested. This research focuses on the utilization of bottom ash (BA) as a sustainable additive to improve the quality of traditional asphalt cement (AC 60/70) in the production of low-cost PAC. The inclusion of BA in AC 60/70 led to a reduction in drain-down and particle loss. Enhanced properties like Marshall stability, ITS, ITMR, ITFL, PD, rut depth, and skid resistance were observed, especially at an optimal BA ratio of 20%. The research highlighted that a 20% BA replacement ratio could reduce the total construction cost of the BA-AC60/70-PAC surface course by 33% compared to conventional PMA-PAC [29].

4 Conclusion and Future Directions

The promising research into recycled and waste materials for pavement construction signifies a major advancement toward sustainable road infrastructure. These alternative materials pave the way for more eco-friendly, affordable, and durable roads. Further assessments of sustainability and real-world viability will enable optimization and wider implementation, aligning with global development goals. Ongoing enhancements in mixture technology and testing in high-traffic settings will realize the immense potential of next-generation sustainable paving.

References

1. Arulrajah A, Ali MMY, Disfani MM, Horpibulsuk S (2014) Recycled-glass blends in pavement base/subbase applications: laboratory and field evaluation. J Mater Civ Eng 26(7)
2. Arulrajah A, Disfani MM, Horpibulsuk S, Suksiripattanapong C, Prongmanee N (2014) Physical properties and shear strength responses of recycled construction and demolition materials in unbound pavement base/subbase applications. Constr Build Mater 58:245–257
3. Barbieri DM, Lou B, Dyke RJ, Wang X, Chen H, Shu B et al (2022) Design and sustainability analyses of road base layers stabilized with traditional and nontraditional additives. J Clean Prod 372
4. Oskooei PR, Mohammadinia A, Arulrajah A, Horpibulsuk S, Emam S (2021) Crushing behavior of recycled waste materials: experimental analysis and DEM simulation. Constr Build Mater 299:124226

5. Naeini M, Mohammadinia A, Arulrajah A, Horpibulsuk S (2021) Stress-dilatancy responses of recovered plastics and demolition waste blends as a construction material. Constr Build Mater 297:123762
6. Senanayake M, Arulrajah A, Maghool F, Horpibulsuk S (2022) Evaluation of rutting resistance and geotechnical properties of cement stabilized recycled glass, brick and concrete triple blends. Transp Geotech 34:100755
7. Arulrajah A, Maghool F, Newman G, Haghighi H, Horpibulsuk S (2022) Cement-treated recycled glass and crushed rock blends: modulus of rupture and stiffness properties. Int J Pavement Eng 23(3):851–861
8. Yaowarat T, Horpibulsuk S, Arulrajah A, Maghool F, Mirzababaei M, Rashid ASA, Chinkulkijniwat A (2022) Cement stabilisation of recycled concrete aggregate modified with polyvinyl alcohol. Int J Pavement Eng 23(2):349–357
9. Sukmak P, Sukmak G, Horpibulsuk S, Kassawat S, Suddeepong A, Arulrajah A (2021) Improved mechanical properties of cement-stabilized soft clay using garnet residues and tire-derived aggregates for subgrade applications. Sustainability 13(21):11692
10. Hoy M, Horpibulsuk S, Arulrajah A (2016) Strength development of recycled asphalt pavement–fly ash geopolymer as a road construction material. Constr Build Mater 117:209–219
11. Hoy M, Horpibulsuk S, Rachan R, Chinkulkijniwat A, Arulrajah A (2016) Recycled asphalt pavement–fly ash geopolymers as a sustainable pavement base material: strength and toxic leaching investigations. Sci Total Environ 573:19–26
12. Hoy M, Rachan R, Horpibulsuk S, Arulrajah A, Mirzababaei M (2017) Effect of wetting–drying cycles on compressive strength and microstructure of recycled asphalt pavement–fly ash geopolymer. Constr Build Mater 144:624–634
13. Xue Y, Arulrajah A, Horpibulsuk S, Chu J (2023) Strength and stiffness performance of geopolymer stabilized washed recycled sands derived from demolition wastes in pavement subgrades. Constr Build Mater 369:130618
14. Yaowarat T, Sudsaynate W, Horpibulsuk S, Chinkulkijniwat A, Arulrajah A, Horpibulsuk J (2021) Mechanical properties of fly ash–asphalt emulsion geopolymer stabilized crushed rock for sustainable pavement base. J Mater Civ Eng 33(9):04021220
15. Arulrajah A, Perera S, Wong YC, Maghool F, Horpibulsuk S (2021) Stabilization of PET plastic-demolition waste blends using fly ash and slag-based geopolymers in light traffic road bases/subbases. Constr Build Mater 284:122809
16. Buritatum A, Suddeepong A, Horpibulsuk S, Udomchai A, Arulrajah A, Mohammadinia A et al (2022) Improvement of tensile properties of cement-stabilized soil using natural rubber latex. J Mater Civ Eng 34(4):04022028
17. Udomchai A, Buritatum A, Suddeepong A, Hoy M, Horpibulsuk S, Arulrajah A, Horpibulsuk J (2022) Evaluation of durability against wetting and drying cycles of cement-natural rubber latex stabilised unpaved road under cyclic tensile loading. Int J Pavement Eng 23(12):4442–4453
18. Nhieu DV, Hoy M, Horpibulsuk S, Karntatam K, Arulrajah A, Horpibulsuk J (2023) Cement–natural rubber latex stabilised recycled concrete aggregate as a pavement base material. Road Mater Pavement Des 24(6):1636–1650
19. Hoy M, Vinh Nhieu D, Suddeepong A, Horpibulsuk S, Arulrajah A, Deng Y, Udomchai A (2023) Stiffness, rutting, and fatigue performance evaluation of cement-natural rubber latex stabilized recycled concrete aggregate. Int J Pavement Eng 24(2):2276204
20. Hoy M, Tran NQ, Suddeepong A, Horpibulsuk S, Buritatum A, Yaowarat T, Arulrajah A (2023) Wetting-drying durability performance of cement-stabilized recycled materials and lateritic soil using natural rubber latex. Constr Build Mater 403:133108
21. Hoy M, Tran NQ, Suddeepong A, Horpibulsuk S, Mobkrathok M, Chinkulkijniwat A, Arulrajah A (2023) Improved fatigue properties of cement-stabilized recycled materials–lateritic soil using natural rubber latex for sustainable pavement applications. Transp Geotech 40:100959
22. Suddeepong A, Hoy M, Nuntasena C, Horpibulsuk S, Kantatham K, Arulrajah A (2022) Evaluation of interface shear strength of natural kenaf geogrid and recycled concrete aggregate for sustainable pavement applications. J Nat Fibers 19(13):6165–6181

23. Maghool F, Arulrajah A, Ghorbani B, Horpibulsuk S (2022) Strength and permanent deformation properties of demolition wastes, glass, and plastics stabilized with foamed bitumen for pavement bases. Constr Build Mater 320:126108
24. Yaowarat T, Suddeepong A, Hoy M, Horpibulsuk S, Takaikaew T, Vichitcholchai N et al (2021) Improvement of flexural strength of concrete pavements using natural rubber latex. Constr Build Mater 282:122704
25. Samingthong W, Hoy M, Ro B, Horpibulsuk S, Yosthasaen T, Suddeepong A et al (2023) Natural rubber latex-modified concrete with PET and crumb rubber aggregate replacements for sustainable rigid pavements. Sustainability 15(19):14147
26. Yaowarat T, Horpibulsuk S, Arulrajah A, Mohammadinia A, Chinkulkijniwat A (2019) Recycled concrete aggregate modified with polyvinyl alcohol and fly ash for concrete pavement applications. J Mater Civ Eng 31(7):04019103
27. Takaikaew T, Hoy M, Horpibulsuk S, Arulrajah A, Mohammadinia A, Horpibulsuk J (2021) Performance improvement of asphalt concretes using fiber reinforcement. Heliyon 7(5)
28. Hoy M, Suddeepong A, Horpibulsuk S, Akkharawongthattana K, Arulrajah A, Buritatum A et al (2024) Improved performance of natural rubber latex-modified asphalt concretes with various types of aggregates. J Mater Civ Eng 36(1):04023495
29. Suddeepong A, Buritatum A, Dasdawan S, Horpibulsuk S, Yaowarat T, Hoy M, Arulrajah A (2023) Mechanical performance of porous asphalt concrete incorporating bottom ash as fine aggregate. J Mater Civ Eng 35(6):04023129

Use of Geosynthetic Materials for Rail Stabling and Maintenance Yard Construction at a Contaminated Site

Udeshini Pathirage, Gabriele Di Marco, Gary Schmertmann, and Jamie McIlquham

Abstract Geosynthetic materials were used extensively in site remediation works prior to construction of a light rail stabling and maintenance yard for the Parramatta Light Rail project, an infrastructure project owned by Transport for New South Wales, Australia. The site was previously a chemical manufacturing premises that had experienced contamination of soil, groundwater, and ground gas, notably including hexavalent chromium (Cr (VI)) and volatile chlorinated hydrocarbon compounds (VCHs). The site also had areas of uncontrolled fill unsuitable for track and associated structure foundations. The geosynthetic materials used in the remediation works included geomembranes, geogrids, and drainage and venting materials. These materials were used in remediation design for containment and isolation of contaminated materials, infiltration reduction, restriction of contaminated groundwater flow, and foundation improvement. The paper describes the remediation design and discusses site requirements, constraints, and design solutions, with emphasis on the use of geosynthetic materials. Remediation works were completed in September 2021.

Keywords Geosynthetic materials · Contamination management · Ground improvement

1 Introduction and Site Conditions

The site is located in Camellia, NSW (see Fig. 1) and has an area of 6 ha. From the 1940s to the 1990s the site operated as a chemical manufacturing facility, resulting in soil, groundwater, and ground gas contamination and areas of uncontrolled filling. This paper concerns site remediation works required prior to construction of stabling and maintenance yard for the Parramatta Light Rail project. The stabling and maintenance yard (SMY) design included entry/exit trackwork, train stabling areas, a train maintenance building, minor structures, and onsite roads/parking.

U. Pathirage (✉) · G. Di Marco · G. Schmertmann · J. McIlquham
WSP Australia Pty Limited, Level 27, 680 George Street, Sydney, NSW 2000, Australia
e-mail: udeshini.pathirage@wsp.com

B. Indraratna and C. Rujikiatkamjorn (eds.), *Recent Advances and Innovative Developments in Transportation Geotechnics*,
https://doi.org/10.1007/978-981-97-8245-1_12

Fig. 1 Site location

Key site conditions considered in the remediation works design are summarised below.

Site ground conditions: Fill overlying natural alluvial soils, with underlying weathered and competent siltstone/sandstone rock at 9–18 m depth. Materials comprise:

- Fill materials: variable in thickness and composition as they have been placed during the working life of the manufacturing facility, with various materials placed under differing levels of engineering supervision and compaction; fill depths up to 12 m on the southern side of the site.
- Alluvial soils: stiff to very stiff clays with interbedded dense sand layers.
- Existing surface elevations ranging from 5 to 7 m AHD prior to remediation.

Groundwater level: Levels prior to remediation are high and within 1 m to 3 m of the existing ground surface. The south bank of the tidally influenced Parramatta River is located approximately 200 m north of the site.

Groundwater Contamination: Site groundwater had contaminants of concern including hexavalent chromium, also known as Cr (VI), and several volatile chlorinated hydrocarbon compounds, or VCHs [1]. Relevant risks of these contaminants to be addressed in the remediation design included:

- Off-site migration of contaminated groundwater (observed prior to remediation).
- Chromium wicking—Cr (VI) in groundwater may be transported upward, above the water table to the ground surface, by the capillary rise of groundwater. This action is referred to as "wicking" (observed prior to remediation).
- Soil vapour generation affecting the SMY site.
- Reduced durability of SMY foundation structures in contact with groundwater.
- Risks exacerbated by high groundwater level and rainfall surface infiltration.

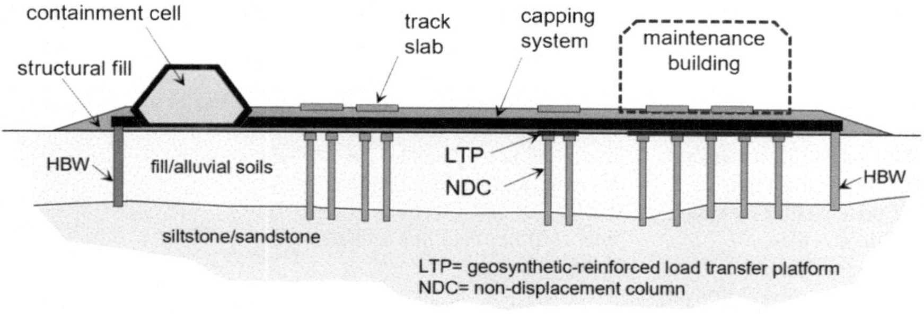

Fig. 2 Indicative cross section of the major components of the remediation design

Soil Contamination: Site soils had contaminants of concern including Cr (VI), VCHs and waste fills. The risk of contaminants causing future environmental harm, being exposed during future operations and harming SMY personnel, or resulting in reduced durability of SMY foundation structures was to be addressed in the remediation design.

2 Design Approach and Geosynthetic Material Use

Remediation design to address the site conditions and risks identified above was completed in a staged process starting with concept design in a Remediation Action Plan [1] and culminating with a detailed design complying with Transport for NSW Authorised Engineering Organisation standards [2–4]. The design was also developed to future SMY construction and operations imperatives and safety constraints. Remediation works construction was to be completed prior to SMY construction.

The major components of the remediation design are illustrated in Fig. 2 and summarised in Table 1.

The remainder of this paper focuses on design and geosynthetic usage for the capping system and ground improvement, with brief discussion of the containment cell.

3 Capping System Design

The required functionality of the capping system is summarised in Table 2. The primary drivers for the use of geosynthetic materials in the capping system were:

Table 1 Remediation design components

Component	Purpose and Function	Geosynthetic Materials
Hydraulic barrier wall (HBW): groundwater cutoff wall along entire 1 km site perimeter; keyed into low-permeability rock with max depth 18 m; soil-bentonite and cement-bentonite slurry wall construction	Restrict groundwater migration; restrict soil vapour migration; note: potential for groundwater level rise inside HBW (i.e. "bathtub effect") unless site has very low rainfall infiltration	Vapour barrier in upper portion of cutoff wall-HDPE Geomembrane
Capping system: multilayer capping system; total cap thickness limited to facilitate positioning above the groundwater level and safe distance below the SMY operating surface; connect to HBW along site perimeter	Restrict surface rainfall infiltration to reduce bathtub effect; restrict chromium wicking; restrict vapour emission; provide gas venting; isolate contaminants from SMY construction and operations personnel; positioned below SMY subsurface services and foundations	Infiltration drainage-geocomposite drainage layer Low-permeability barrier-coated GCL Top/bottom of capillary break layer: separation geotextile
Ground improvement: non-displacement concrete ground improvement columns; enlarged column heads; load transfer platforms (LTPs) above column heads in some areas	Provide for suitable track and building structure foundations; provide suitable durability for site exposure conditions; positioned below capping system (no penetration)	Geosynthetic reinforcement (geogrid) in LTPs
Containment cell: fully lined and capped cell for placement of contaminated soils excavated during site remediation works	On-site encapsulation of contaminated soils; Mound configuration located outside of SMY operations footprint	Base liner and capping system: (a) low-permeability barriers-coated GCLs; (b) Leachate/infiltration drainage, gas/vapour venting, and capillary break-geocomposite drainage layers; and (c) geosynthetic reinforcement (geogrid) in cover soil

- The reduced thickness of geosynthetic materials, compared to compacted clay barriers and gravel drainage layers, allowed for the capping system to be positioned generally above the groundwater level while remaining below SMY subsurface services and foundations.
- The very low permeability of a geosynthetic capping system, compared to compacted clay barriers, would restrict rainfall infiltration and potential for groundwater level rise within the HBW-surrounded site (i.e. reduce "bathtub

Table 2 Design functions of each cap component

Component	Objectives addressed[1]					Primary functions	Secondary functions
	VE	WI	CW	PR	IS		
Structural fill	X	X		X	X	Working surface for future site constructions	Surface water drainage, isolation
Infiltration drainage layer (Geocomposite drainage layer)		X		X		Drainage of water infiltrating through structural fill	Protect underlying C-GCL
Liner (Coated GCL)	X	X			X	Barrier to water infiltration	Barrier to vapour emission
Bearing soil	X	X		X	X	Surface for C-GCL placement	Enhance infiltration resistance
Capillary break layer	X		X		X	Chromium wicking barrier	Vapour collection using HDPE venting pipes
Separation geotextile			X	X		Filter to prevent fines entering Cap. break	Visual marker layer
Additional for cap type 2							
Stabilised sand				X	X	Future site use requirement	Protects layers beneath
Cushion geotextile				X		Protection to liner	
Liner (2 × HDPE geomembrane)	X	X	X			Barrier to water infiltration to keep Cap. break dewatered	Vapour emission barrier; wicking barrier

Note 1 VE = Vapour Emission; WI = Water Infiltration; CW = Chromium Wicking; PR = Cap Protection; IS = isolate/protect SMY personnel from contaminants

effect"). Reducing the bathtub effect was a significant factor for the management of the risk of Cr (VI) wicking. This is because the primary anti-wicking feature in the capping system is a thick blanket layer of drainage aggregate that requires a dewatered state to function effectively.

As part of the capping design to mitigate geosynthetic material vulnerability to damage, long-term site operations procedures were developed to reduce risk of inadvertent penetration or disruption of the capping system during future SMY construction or excavation works.

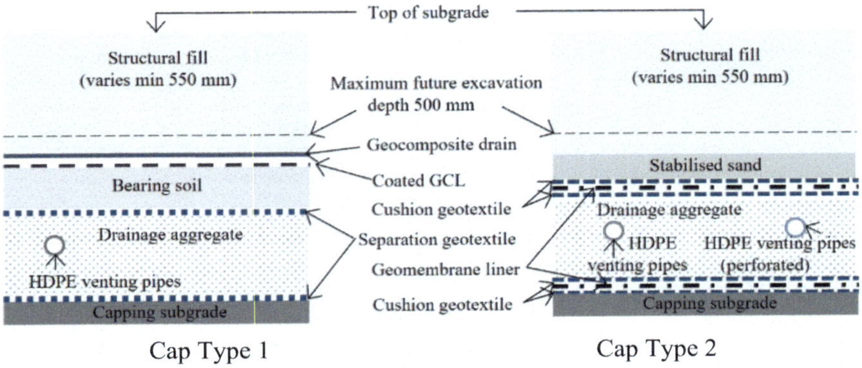

Fig. 3 Capping system details

The capping system consists of several layers: a low-permeability liner and overlying subsurface drainage, a capillary break layer, passive vapour ventilation infrastructure and a vapour barrier for buildings. As indicated above, the capillary break layer is a drainage aggregate blanket (400 mm thick) that functions to restrict Cr (VI) wicking. The vapour collection system was incorporated with the capillary break layer.

Cap type 1 (Fig. 3) was to be constructed over the entire site within the boundary of the HBW, except for beneath SMY buildings. The primary objectives of Cap type 1 are to reduce water infiltration, reduce vapour emissions and chromium wicking, and isolate SMY construction and operations personnel from exposure to contaminants. Cap type 2 (Fig. 3) has similar functionality but is constructed beneath the maintenance building and includes a stand-alone vapour management system with two geomembrane vapour barrier layers [2].

Table 2 describes the design features and processes of each component of the capping system, including the geosynthetic materials.

4 Ground Improvement Design

4.1 Geotechnical Model

Subsurface conditions and stratigraphy are summarised in Sect. 1, noting that ground improvement is particularly relevant to previously filled areas. Fill materials are variable in thickness and composition as they have been placed during and after the working life of the facility, with varying materials placed under differing levels of engineering supervision. This includes areas in the southern part of the site that have been previously remediated to depths of up to 12 m using excavate and replace methods.

Groundwater is present at relatively shallow depth of approximately 1 to 3 m below the existing surface level. The contaminants present in soil and groundwater were considered in durability design for the materials used for the ground improvement (GI).

4.2 Design Requirements

The key design requirements for GI were to achieve acceptable total and differential settlement requirements for future stabling yard infrastructure, including concrete track slabs and maintenance and storage structures. Foundation loading and settlement requirements were as follows:

- Track slab/hardstand area.
- Surface loading of 40 kPa.
- Light rail vehicle loading comprising 8 × 12.5 tonne axle loads.
- Total residual settlement < 50 mm and differential settlement < 0.3%.
- Building structures.
- Surface loading of 40 to 60 kPa.
- Total residual settlement < 20 mm and differential settlement < 0.1%.
- Landscaped areas.
- Surface loading of 25 kPa.
- Total settlement < 200 mm and differential settlement < 0.5%.

An important factor was that the GI design was required to be compatible with the capping system adopted for the project (Sect. 3). A key consideration in this respect was that no penetrations of the capping layers could be made, meaning that conventional piled foundations could not be used. In addition, the GI system could not induce excessive local displacement/strain in the capping system. Another design factor was that the hydraulic barrier wall (HBW) described in Sect. 2 was constructed as the first step in the remediation works and it was necessary that the GI construction process did not damage the installed HBW through displacement or pore pressure effects. Table 3 provides a summary of advantages and disadvantages of considered GI methods listed as ranked from top to bottom.

4.3 Selected GI Design

The GI design process involved review of the geotechnical model of the site and adopting different improvement techniques, depending on the conditions present and based on initial one-dimensional settlement calculations. In the northern part of the site, the existing fill layer was relatively thin and therefore conventional proof rolling techniques were adopted for GI.

Table 3 Advantages and disadvantages of each GI option

GI option	Advantages	Disadvantages
NDCs	• Quick (estimate 3 months on site) • Existing capping fill may be able to act as LTP • Designed to restrict settlement at underside of proposed concrete track slab • Short inclusions–low risk of creating pathway for contaminant migration to bedrock • Lower risk of refusing in very stiff residual soil than drill displacement columns (DDCs)	• Requires spoil disposal • Concrete and geofabric durability would need to be assessed
DDCs	• Quick (estimate 3 months on site) • Existing capping fill may be able to act as LTP • Designed to restrict settlement at underside of proposed concrete track slab • Short inclusions–low risk of creating pathway for contaminant migration to bedrock	• May require non-displacement rig if clay is very stiff–requires spoil disposal • Concrete and geofabric durability would need to be assessed • Risk of early refusal on obstructions • Ground movements caused by method impacting HBW or adjacent GI columns
Jet grouting	• Relatively slow (estimate 6 months on site) • Relatively low volume of spoil generated, which is already stabilised with cement • Only the compost layer is targeted for treatment	• Significant health and safety controls required, high-pressure air and water used in method higher risk to workers undertaking work • Concrete durability would need to be assessed • Uncertainty on size of columns until site trial completed, could lead to increase in cost and programme duration for work

At the southern end of the site, fill thicknesses increased, including previously backfilled remediation pits up to 12 m deep. GI was required in these areas to achieve design requirements. Different ground improvement methods were considered at the commencement of design, including ground improvement columns (displacement and non-displacement methods), soil mixing, jet grouting, surcharge, and excavate-replace. The design assumptions for column capacity used published design methods for drilled shafts to estimate likely shaft and base resistance. These assumptions were tested and verified on site during construction using a combination of dynamic and static load testing and longer columns were required in areas with more stringent settlement criteria.

After assessment of the project constraints, advantages/disadvantages of each approach, and consultation with stakeholders and a specialist ground improvement

Fig. 4 **a** 3D model layout of NDC columns, **b** photograph after installation of the NDC columns

contractor, the use of non-displacement columns (NDCs) comprised of unreinforced concrete was selected as the preferred GI method. Typical design NDC column spacing was 2 m. Figure 4 shows the layout of NDC columns in a 3D model and a photograph of the installed NDC columns at the site.

To improve settlement performance, the design adopted enlarged heads at the top of the NDC columns, with the top of the heads positioned below the bottom of the site capping system. The enlarged heads reduced the span distance between columns, improving settlement performance of the system. A high-strength geosynthetic reinforced load transfer platform (LTP) was adopted below the building raft slabs and some of the track slab areas to enhance foundation load distribution and achieve the required settlement performance. The use of geosynthetic reinforcement allowed LTP thickness to be reduced to allow the NDC column heads to extend upward as close as possible to the site capping layer.

Studies were undertaken during design to assess durability considerations for the concrete used for the ground improvement columns and the geosynthetic material used in the LTP under exposure to site contaminants.

5 Containment Cell Design

The containment cell (CC) was designed with a mound configuration to contain excavated contaminated material from cut and fill earthworks during the remediation works. The CC has a continuous base liner and capping system including geosynthetic components as described below.

- The CC base liner system has a low-permeability liner system that includes a coated geosynthetic clay liner (C-GCL) to minimise downward infiltration of CC leachate. This liner is overlain by a geocomposite drainage layer for leachate collection and drainage to extraction sumps.

- The base of the capping system comprises a low-permeability liner that has a C-GCL and a 200 mm thick layer of bearing soil to reduce downward infiltration. Below these materials is a geocomposite drainage layer to collect gas/vapour and restrict Cr (VI) wicking.
- Above the C-GCL in the capping system is another geocomposite drainage layer to collect and drain surface infiltration and keeps the cover soil dry. Within the cover soils above these materials, the design includes geosynthetic reinforcement (geogrid) to provide suitable capping system veneer stability.

The reduced thickness of geosynthetic materials, compared to compacted clay barriers and gravel drainage layers, allowed for greater CC storage capacity to be realised within a given landform geometry. As part of the design, long-term site operations procedures were developed to reduce risk of inadvertent penetration or disruption of the CC geosynthetic materials during future SMY construction/excavation works.

6 Summary

This paper describes the remediation of a contaminated site for the future use of a stabling and maintenance yard (SMY) for Parramatta Light Rail project. The design included a hydraulic barrier wall, an integrated capping system, and ground improvement with the use of non-displacement columns (NDCs) to address the following design challenges:

- Fill overlying natural alluvial soils, with underlying weathered and competent siltstone/sandstone rock.
- Off-site migration of contaminated groundwater.
- Chromium wicking–possible upward transportation of Cr (VI) in groundwater to the ground surface, by the capillary rise of groundwater.
- Soil vapour generation affecting the SMY site.
- Reduced durability of SMY foundation structures in contact with groundwater.
- Risks exacerbated by high groundwater level and rainfall surface infiltration.

The design of the integrated capping system included geosynthetic materials for the following advantages.

- The reduced thickness of geosynthetic materials, compared to traditional compacted clay barriers and gravel drainage layers, allowed for the capping system to be positioned generally above the groundwater level while remaining below SMY subsurface services and foundations.
- The very low permeability of a geosynthetic capping system, compared to compacted clay barriers, would restrict rainfall infiltration and potential for groundwater level rise within the HBW-surrounded site (i.e. reduce "bathtub effect").

Acknowledgements Authors acknowledge WSP/Golder personnel involved in various parts of the project from design to construction in geo-environmental, geotechnical, and contaminated land remediation teams. Also, we acknowledge Transport for NSW for granting permission to showcase the remediation design presented herein and WSP client, Ventia Pty Ltd.

References

1. WSP/Golder Associates Pty Ltd, remediation action plan (RAP) for 6–10 Grand Avenue, Camellia, NSW, reference: 1775480-006-R-Rev0 (2017)
2. WSP/Golder Associates Pty Ltd, integrated capping system detailed design report, reference: 1898648-1200-R-Rev0 (2020)
3. WSP/Golder Associates Pty Ltd, ground improvement detailed design report, reference: 1898648-1300-R-Rev0 (2020)
4. WSP/Golder Associates Pty Ltd, containment cell detailed design report, reference: 1898648-1100-R-Rev0 (2020)

Significance of Rock Slopes Monitoring for the Protection of Transportation Networks and Recreational Coastal Areas

Anna Giacomini, Abigail Watman, Davide E. Guccione, Farshad Bahootoroody, and Klaus Thoeni

Abstract Rock slope instabilities, such as rockfalls and rockslides, can threaten human life along major transportation networks and recreation coastal areas, affect safety during infrastructure construction and significantly interrupt transportation along highways and railways with substantial economic losses. Accurate predictions of the associated risks and their effective reduction require a thorough collection of rock mass structural conditions and triggering events' evolution with time to infer appropriate modelling parameters for the design and implementation of the most suitable mitigation measures. Close-range terrestrial photogrammetry and Unmanned Aerial Vehicles (UAVs) have been used to monitor hazardous and hardly accessible rock slopes located above popular recreational areas along the Newcastle coastline in NSW. Valuable inventories of rock mass structural data and rockfall spatial–temporal occurrence and magnitude have been collected to identify major structures affecting rock slope stability and sources of rock instabilities (detachments) over extended periods of time. Their volume and frequency have been accurately estimated. Results highlight the crucial role played by the identification of appropriate input parameters to adequately simulate realistic rockfall scenarios and the significance of the frequency of data collection for accurate predictive modelling.

Keywords Proximity remote sensing · Slope monitoring · Rockfall hazard · Transportation protection design

---

A. Giacomini (✉) · A. Watman · D. E. Guccione · F. Bahootoroody · K. Thoeni
Centre for Geotechnical Science and Engineering, University of Newcastle, Callaghan, NSW 2308, Australia
e-mail: Anna.Giacomini@newcastle.edu.au

B. Indraratna and C. Rujikiatkamjorn (eds.), *Recent Advances and Innovative Developments in Transportation Geotechnics*,
https://doi.org/10.1007/978-981-97-8245-1_13

1 Introduction

Every year in Australia, millions of dollars are spent on geotechnical assessment, stabilization and mitigation measures to manage the risk associated with major natural hazards such as rockfalls and other types of rock instabilities. These phenomena threaten life, pivotal national transportation networks and numerous recreation areas along the densely populated coastal fringes of Australia. A predicted increase in the intensity of extreme weather events may further impact the rate and severity of rock cliff hazards in the coming years [1] as strong weather patterns will lead to more frequent and stronger storms, with heavy rain, strong winds and powerful coastal wave conditions affecting the rock mass structure and result in an increase of likelihood of rock instability hazards. An accurate quantification of the risks associated with rock instabilities along affected transportation networks and popular recreational scenic rock cliffs is crucial for the prioritization of management and mitigation measures and the associated allocation of resources.

Recent developments of close-range proximity remote sensing technologies for slope monitoring, such as terrestrial laser scanners (TLS), photogrammetry, airborne Light Detection and Ranging (LiDAR), combined with Unmanned Aerial Vehicles (UAV), permit high-resolution data acquisition, often at low cost, in highly inaccessibility areas. Pivotal structural, spatial and temporal data to quantify the frequency of rock slope failures and their triggering mechanisms can be collected relatively quickly and integrated at minimal post-processing costs. 3D high-resolution Digital Surface Models (DSM) allow for the mapping of geostructural features and identification of in situ block distributions to assess the probability of occurrence in a Quantitative Risk Assessment (QRA) framework. The comparison of DSM from different observation periods allows for the identification of geometrical changes (change detection) and quantification of detached volumes. However, an accurate automatic generation and reliable comparison of DSM taken at different times to quantify the evolution of the cliffs remain an open issue and the identification of the required accuracy and of the appropriate processing parameters for accurate predictions is not straightforward.

This study focuses on a sub-vertical coastal rock cliff located at South Newcastle Beach in NSW. Proximity remote sensing techniques, structural analyses and change detection investigations have been combined to investigate the slope stability and provide valuable insight for risk mitigation management. Over the past decades, this cliff has experienced recurrent rockfalls, leading to multiple examinations of slope stability and the implementation of measures to mitigate the associated rockfall risks.

2 Methodology

Photogrammetric imagery of the exposed rock face is acquired during regular aerial surveys conducted via a DJI Phantom 4 Pro RTK (20 MP camera). Three-dimensional (3D) point cloud models of the rock face are then generated at each survey period by post-processing the collected images in Agisoft Metashape [2] and CloudCompare [3] to assess the most appropriate point density for structural mapping and conduct further change detection analyses to determine the volumes of rocks detached from the rock faces between each monitoring period. The open-source software package CloudCompare is used for this purpose: the volumes of the blocks are estimated by using a "cloud-to-cloud distance" computation analysis for which the areas of detached rocks are firstly estimated by comparing two sets of point clouds collected at different periods, and the "distance" between two point clouds is then integrated over these areas to estimate the volumes by a user-defined Python script. The libraries *meshio* and *scipy* are used to convert the point clouds into meshes (closed objects). Figure 1 shows the "step by step" approach followed in the study for which "cloud-to-cloud" distances are displayed in the form of a contour plot of computed distances for each of the points within the compared models collected at different periods. High-resolution photographs are also compared to verify the correctness of the observed detachments and their locations on the rock face.

Estimated size and frequency data of rocks detached from the rock face between surveyed periods are considered for further hazard and risk assessments and mitigation measures design. For this purpose, monitoring results are used to conduct site-specific rockfall analyses via RocFall3 [4] and predict expected kinetic energies and run-out distances of the falling blocks at the toe of the rock faces, representing an important potential risk for infrastructure and human activities along the base of the rock cliff.

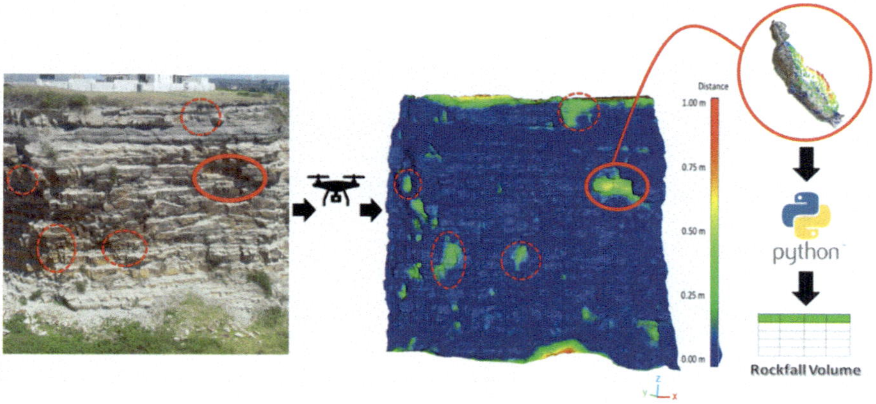

Fig. 1 Conceptual framework of the data collection and processing

3 Case Study Application

This paper examines, as an example, the case study of a sedimentary coastal rock cliff situated along a heavily frequented segment of the Bathers Way coastal walkway at Newcastle South Beach (NSW). The rock cliff spans approximately 300 m in length and reaches a height of 40 m at its highest point. The walkway, originally part of the Shortland Esplanade roadway between Newcastle Beach and King Edward Park, underwent significant changes over the last two decades after a series of rock instabilities recorded since the early '90s: several investigations for mitigating the risk of rockfall were conducted with the goal of re-establishing safe pedestrian access. The cliff was stabilized with around 200 fully grouted rock bolts, localized excavation and the construction of a 90 kJ capacity, 1.8 m-high rockfall protection fence between the cliff and walkway in 2005 [5], further increased in height to 3.2 m along the northern end of the site between 2016 and 2017 during the renovation works of the walkway and sea wall in the site area [6]. As part of the ongoing South Newcastle Beach Bathers Way coastal revitalization project, the site is currently under construction with plans for further updating the walkway and building a skate park at South Newcastle Beach (see Fig. 2).

While current mitigation measures in place, namely the rockfall fence and the rock bolting, have resulted effective in intercepting the rock falling from the cliff and preventing the detachment of larger blocks, further monitoring of the rockface stability conditions has been conducted over the last 12 months to assess residual hazards and support a robust risk management plan in view of the intense pedestrian activity expected in the coming years along the renovated areas located at the bottom of the cliff.

Fig. 2 South Newcastle Beach sub-vertical cliff face with indicated existing rockfall protection fence and ongoing coastal revitalization works (modified after [6])

A series of close-range photogrammetric surveys have been conducted along the coastal cliff. 3D point cloud data obtained from drone imagery over multiple periods have been used to assess the rock cliff's structural features and to identify the source location of unstable rocks and their volume. 3D models of the rock face have been built in Agisoft Metashape and optimized using CloudCompare and Meshlab [7]. Thorough structural and change detection analyses have been conducted to identify the principal geostructural features of the exposed rock mass and assess detachments from the rock face. Survey data have then been considered to conduct rockfall 3D analyses by importing a representative 3D mesh of the rock face (from the drone survey of June 2023) in RocFall3. In the model, different regions have been assigned to the corresponding rock material layers named sandstone, siltstone, talus cover/tuff and concrete (representing the walking path). The coal seam layer was not simulated given its very low thickness compared to the other material layers and the challenges associated with the setting of horizontal layers in the software.

Input parameters such as coefficients of restitution and friction angle have been chosen according to Giacomini et al. [8] (siltstone and sandstone) and RocFall3 material library (talus cover/tuff and concrete) (see Table 1).

To assess the rock face predisposition to rockfall, the existing rockfall fence was removed from the model. Simulations have considered rock size and material identified by the survey, estimated during the change detection analysis and regular walk path surveys conducted over the last 12 months: the maximum block size estimated by change detection within the layer of siltstone rock material over a recent two-month period (November 2022–January 2023) and the maximum detachment size from the sandstone rock layer observed along the toe of the slope have been considered [6]. For the siltstone layer, we have assumed that the block did not fracture along its fall or before reaching the bottom of the cliff, reducing, therefore its size at impact at the bottom of the cliff.

The analysis was conducted by using both "lumped mass" and "rigid body" approaches. Rock shapes considered in the latter approach have been chosen according to the shape of blocks resulting from the survey investigation. In the simulations, a normal scaling factor by velocity and the sampling technique Monte Carlo method have been considered. Line seeders located on top of both the siltstone and sandstone layers have been used to represent the possible blocks' detachment

Table 1 Input material parameters used in RocFall3 analyses

Name	Colour	R_n			R_t			Friction angle (°)
		Value	Min	Max	Value	Min	Max	
Sandstone	Orange	1	0.82	1	0.4	0.25	0.55	25
Siltstone	Light grey	0.5	0.29	0.71	0.7	0.49	0.91	26
Talus/tuff	Light blue	0.32	0.17	0.47	0.8	0.62	0.98	30
Concrete	Grey	0.48	0	1	0.53	0.02	1	41.6

locations in the simulations (1000 blocks per seeder have been released). A 0.5 m/s (± 0.3 m/s) translational velocity and 90°/s (± 15°/s) rotational velocity were used as initial conditions of the seeders. Energies and run-out distance of the block (walk path potential reach) have been assessed.

4 Results

Change detection analysis has allowed us to estimate around 50 rock detachments from the rock face over a two-month period, with most detachments of different sizes originating from the siltstone (38 blocks) and sandstone (6 blocks) rock layers for a total volume of rock detached of almost 1.5 m^3 (see Table 2). The maximum siltstone size block was estimated over 0.8 m^3 with a flat rectangular shape (1930 kg).

The analysis has also permitted us to identify the detachment source location of each block within the different rock material layers (see Fig. 3).

While several potentially unstable sandstone blocks of significant size have been identified on the rock face during the survey, their detachment is still prevented by the presence of rock bolting. According to the size and shape of blocks observed during

Table 2 Rockfall volume summary statistics by material layer

Material	# blocks	Volume (m^3)			
		Min	Max	Mean	Total
Sandstone	6	0.0024	0.1013	0.0261	0.1568
Siltstone	38	0.0007	0.8219	0.0322	1.2231
Talus cover/tuff	3	0.0107	0.0451	0.0266	0.0797

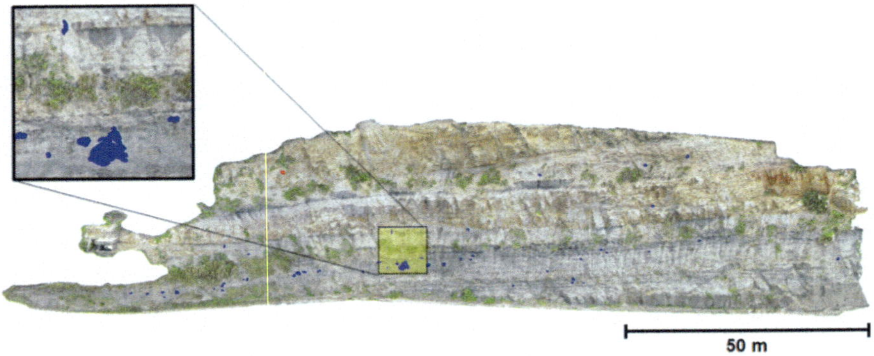

Fig. 3 Change detection results on the 3D model of the rock face: coloured areas indicating the location of detached blocks on the rockface, red detachments corresponding to the first monitoring period and blue areas corresponding to the second monitoring period

Fig. 4 Sandstone block of 235 kg observed between April 2022 and January 2023 at the bottom of the cliff and contained by the existing rockfall barrier

the monitoring activity, lumped mass and rigid body RocFall3 analyses have been conducted to assess the rockfall hazard at the bottom of the rock cliff and along the pathway. For a matter of space limit, only rigid body results are reported. Siltstone blocks of 1930 kg ($\pm$ 390 kg) with extruded polygon rectangle (2:3) and extruded polygon rectangle (1:2) shapes have been used in the simulations. For the sandstone layer, the mass of the biggest fallen block observed between April 2022 and January 2023 at the toe of the cliff (235 kg $\pm$ 60 kg block with a density of 2347 kg/m^3) was considered (see Fig. 4). The block was simulated by extruded polygon rectangle (2:3) and cube shapes.

Figures 5 and 6 show a summary of the rockfall rigid body modelling results in terms of kinetic energy and simulated rockfall trajectories and run-out distances at the base of the rock cliff, respectively.

Results of the simulations conducted without considering the presence of the current rockfall barrier show that over 50% of the blocks reach the walkway with up to 30% of the blocks reaching the edge of the walking path. The installation of a rockfall barrier remains, therefore, a crucial mitigation measure for the impeding rockfall hazard. Most of the small sandstone blocks observed during the monitoring activity could be easily contained by a 90 kJ capacity barrier.

However, the simulation of large blocks along the siltstone layer evidence energies of up to 200 kJ, particularly for simulated trajectories at the bottom of the cliff within areas currently under renovation works, highlighting the need for ongoing monitoring to support robust rockfall risk assessment and management.

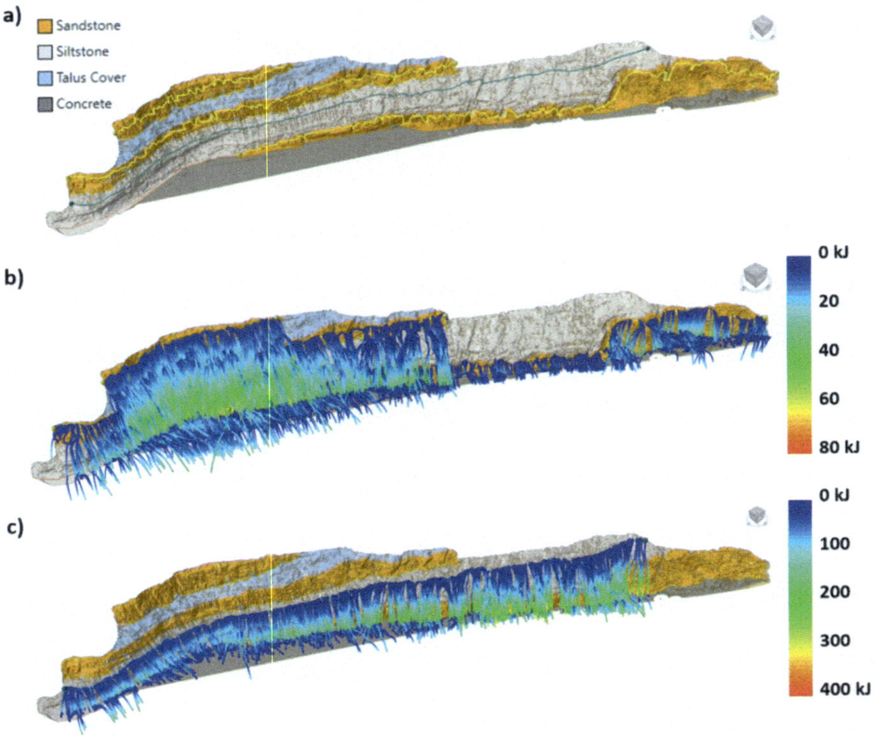

Fig. 5 Rockfall modelling results: **a** rock material layers assigned to the Rockfall3D model; **b** kinetic energy results for the block falling from the sandstone material layer; **c** kinetic energy results for the block falling from the siltstone material layer

5 Conclusions

The paper presents the application of an integrated approach to assess the stability of a sub-vertical rock cliff exposed to potential rockfall hazards. Proximity remote sensing techniques have been combined with change detection analysis and three-dimensional rockfall trajectory modelling to collect size and frequency data of detached blocks over a two-month monitoring period. Monitoring results have been used to infer modelling input parameters for the layered sedimentary rock face at South Newcastle Beach. Results highlighted the importance of conducting a thorough monitoring campaign to infer robust parameters for rockfall modelling and design, thereby increasing the certainty of rockfall hazard assessments and future quantitative risk management.

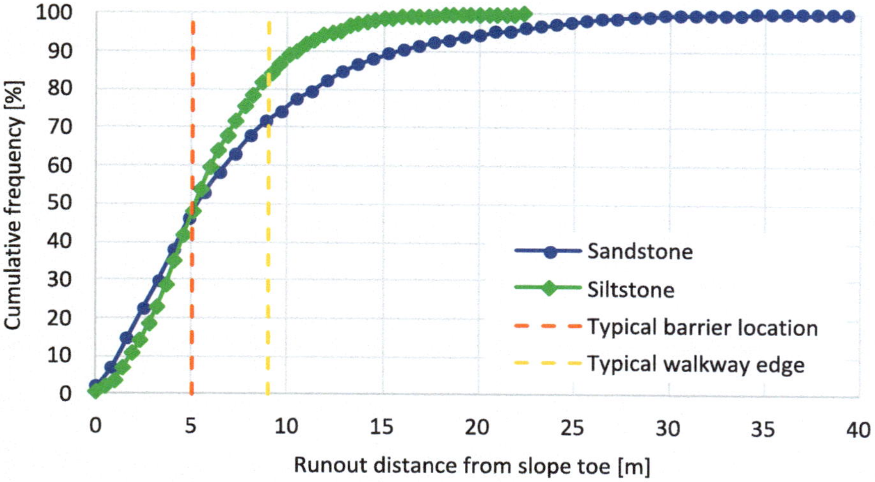

Fig. 6 Run-out distances estimated at the bottom of the cliff for sandstone and siltstone blocks. The location of the existing rockfall barrier and the walkway's edge are also indicated by red and yellow dashed lines, respectively

Acknowledgements This research was funded by the Australian Research Council, grant number DP210101122. The authors would like to acknowledge Merrick Jones (Tetra Tech) and the Newcastle City Council for providing access to the site to collect aerial imagery.

References

1. Middelmann MH (ed) (2007) Natural hazards in Australia. Identifying risk analysis requirements. Geoscience Australia, Canberra
2. Agisoft (2021) Metashape (1.8). [Software]
3. Girardeau-Montaut D (2020) CloudCompare (2.11.3). [Software]
4. Rocscience. RocFall3 (1.012). [Software]. https://www.rocscience.com/. Last access 11 Dec 2023
5. Kotze G (2011) Shortland Esplanade, South Newcastle Beach, Australia—a case study in rockfall risk management. Austral Geomech 46(2):71–88
6. Guccione DE, Watman A, Theoni K, Buzzi O, Giacomini A (2023) Coastal cliffs rockfall analyses and mitigation measures assessment using RocFall3: a case study along Shortland Esplanade in Newcastle, NSW (Australia). In: RocScience international conference (RIC2023) proceedings, pp 827–834
7. Cignoni P, Calleri M, Corsini M, Dellepiane M, Ganovelli F, Ranzuglia G (2008) MeshLab: an open-source 3D mesh processing system. In: 6th Eurographics Italian chapter conference, pp 129–136
8. Giacomini A, Thoeni K, Lambert C, Booth S, Sloan SW (2012) Experimental study on rockfall drapery systems for open pit highwalls. Int J Rock Mech Min Sci 56:171–181

Liquefaction Countermeasure for Residential Buildings Using Gravel-Mixed Tire Chips and Its Validation Through an Integrated Approach

Hemanta Hazarika, Yutao Hu, Gopal Santana Phani Madabhushi, and Stuart Kenneth Haigh

Abstract The vertical drain method, widely used to prevent earthquake-induced liquefaction, utilizes vertical drains to dissipate excess pore water pressure and reduce liquefaction risk. Hazarika et al. (Soils Found 60:315–326, 2020 [11]) developed a cost-effective technique to shield residential buildings from such damage, using a mixture of tire chips and gravel (GTCM) under foundations. This approach also shows promise as a drainage enhancer in vertical drains. Incorporating GTCM in vertical drains, the research involved small-scale shaking table tests, two-dimensional simulations, and cyclic undrained triaxial tests to validate its efficacy. The findings indicated that GTCM drains effectively prevent foundation soil liquefaction, with optimal results observed at a gravel content of 50–60% by volume, striking a balance between controlling excess pore water pressure and maintaining drain stiffness.

Keywords GTCM · Liquefaction · Model tests · Finite elements

1 Introduction

Over recent years, there has been a global surge in interest regarding waste tire-derived geomaterials (WTGM) as innovative solutions for ground improvement. Notably, Japan has emerged as a leader in this field, with significant progress in both research and practical applications beginning in the early 2000s [19]. This period

H. Hazarika (✉) · Y. Hu
Kyushu University, Fukuoka 819-0395, Japan
e-mail: hazarika@civil.kyushu-u.ac.jp

G. S. P. Madabhushi · S. K. Haigh
University of Cambridge, Cambridge CB2 1PZ, UK

B. Indraratna and C. Rujikiatkamjorn (eds.), *Recent Advances and Innovative Developments in Transportation Geotechnics*,
https://doi.org/10.1007/978-981-97-8245-1_14

saw the development of numerous groundbreaking recycling methods of waste tires for civil engineering uses [7–9, 11, 13, 27, 28]. Hazarika and Yasuhara [12] provide an extensive compilation of global studies and practical implementations involving waste tires.

The utilization of WTGM falls into two primary categories: (1) stand-alone usage, where the materials are employed in their original form or as processed elements like tire shreds and chips, without blending with other geomaterials. This approach is exemplified in the construction of retaining walls, temporary dams, waste disposal facility embankments, and coastal protections against tsunami damage using whole tires [8, 10, 28]. Karmokar [18] detailed a project in Hokkaido, Japan, which enhanced highway embankment drainage using tire shreds. (2) Composite application, which involves mixing WTGM with soil or other substances. A notable instance of this method is the use of a mixture of tire chips and gravel (GTCM) as a horizontal inclusion under the foundation of residential buildings to prevent vibration-induced and liquefaction-induced damage during earthquakes [11]. The adoption of WTGM in civil and geotechnical engineering is driven by their advantageous properties, such as being lightweight, vibration-dampening, highly permeable, and thermally conductive.

The vertical drain method, a recognized liquefaction countermeasure in practice, primarily aims to mitigate excess pore water pressure during earthquakes, thus preventing liquefaction [1]. Garcia-Torres and Madabhushi [5] have explored the effectiveness of these drains under structures. Given the successful application of tire chips and gravel mixture (GTCM) in horizontal reinforcing inclusions, it seems feasible to also use GTCM in the vertical drain method for similar purposes. A new sustainable technique was therefore proposed, by using GTCM as prefabricated drainage system to mitigate liquefaction [14].

In this research, an integrated approach comprising of model tests, element tests, and numerical simulations was adopted to validate the technique developed. First, cyclic undrained triaxial tests were performed to evaluate the liquefaction susceptibility of GTCM. Then, a series of small-scale 1-g model shaking table tests was performed, to evaluate the technique's effectiveness. In addition, a prototype two-dimensional simulation of the technique was performed.

2 Dynamics Properties of GTCM

In this test series, the large cyclic triaxial testing apparatus was utilized. Consolidated undrained (CU) tests were conducted on specimens measuring 100 mm in diameter and 200 mm in height. These tests aimed to assess the liquefaction resistance and large strain shear modulus of GTCM.

2.1 Materials and Test Procedures

Figure 1 presents the particle size distribution curves for both the gravel and tire chips (TC) used in the study. Additionally, Table 1 lists other physical properties of these materials, including the maximum diameter of the gravel and tire chip particles (D_{max}), the coefficient of curvature (C_c), and the coefficient of uniformity (C_u). To minimize the influence of sample size on the experimental results, the maximum grain size of both the TC and the gravel was restricted to less than one-sixth of the specimen diameter. According to the standards of the Japanese Geotechnical Society (JGS 0131), the gravel used in this study is classified as poorly graded (SP). The specific gravities (G_s) of the gravel and TC, determined following the guidelines of the Japanese Geotechnical Society (JGS 0111) [17], were 2.81 for the gravel and 1.17 for the TC, respectively.

An isotropic consolidation pressure was exerted on the samples with different volumetric gravel fractions (Gf). A constant initial back pressure was maintained at 200 kN/m^2. The samples underwent consolidation at effective confining pressures of 50 and 100 kN/m^2. Following this, a sinusoidal cyclic axial loading was applied

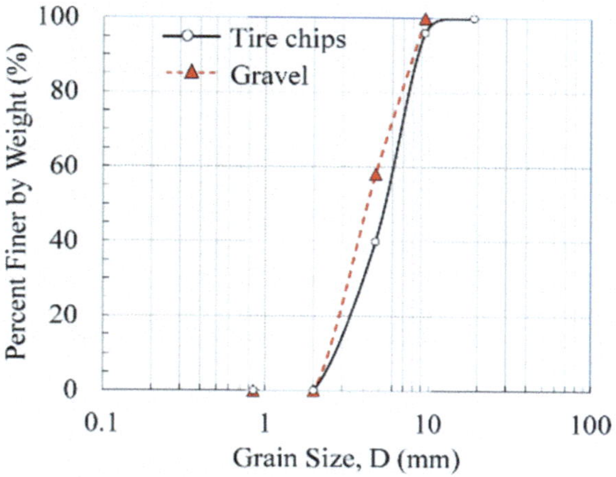

Fig. 1 Particle size distribution curves of gravel and tire chips

Table 1 Physical properties of the materials

Material	Gravel	Tire chips
Specific gravity, G_s	2.81	1.17
Maximum void ratio, e_{max}	0.829	1.183
Minimum void ratio, e_{min}	0.562	0.966
Maximum particle size, D_{max} (mm)	9.5	12.0
Mean particle size, D_{50} (mm)	4.2	5.3
Coefficient of curvature, C_c	2.17	2.32
Coefficient of uniformity, C_u	0.90	0.94

at a frequency of 0.1 Hz under undrained conditions. All the tests were carried out on isotropically consolidated samples set at a relative density of 50%, and they were subjected to various cyclic stress ratios ($\sigma_d/2\sigma'_{3c}$).

2.2 Results and Discussion

Figure 2 illustrates the impact of confining pressures on the generation of pore water in GTCM with a Gf of 30% and a relative density (D_r) of 50%, at a cyclic stress ratio of 0.35. The GTCM samples demonstrated increased resistance to liquefaction, even under lower confining pressures, a behavior akin to that observed in denser sand samples subjected to higher confining pressures. Notably, a reduction in effective confining pressure significantly boosted the liquefaction resistance of the GTCM samples with a Gf of 30%. This enhancement in resistance could be attributed to the decrease in hydraulic conductivity of the mixture as the effective confining pressure decreases, as suggested by Edil and Bosscher [3].

Figure 3 depicts the maximum excess pore water pressure ratio ($R_u = u/\sigma'_{3c}$) for samples with a relative density (D_r) of 50%. These data are plotted at a stress ratio of 0.3, considering different gravel fractions (Gf) for a fixed number of cycles ($N = 20$). Since all the testing conditions, including relative density (D_r), loading frequency, cyclic stress ratio ($\sigma_d/2\sigma'_{3c}$), and the number of the cycles, were the same for the GTCM samples with different gravel fractions, the same input energy was applied to the samples. Consequently, a lower maximum excess pore water pressure indicates higher liquefaction resistance. The data reveal that liquefaction resistance decreases as the Gf drops from 100 to 87%, but then increases with any further reduction in the gravel fraction. In pure gravel samples (Gf = 100%), the medium-dense state and significant dilative behavior contribute to high liquefaction resistance. Adding a small amount of tire chips alters the matrix composition; while

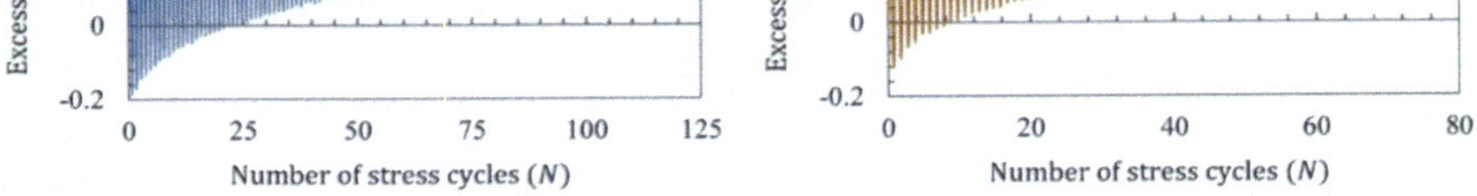

Fig. 2 Influence of confining pressure on the generation of excess pore water pressure: **a** $\sigma'_{3c} = 50$ kN/m^2; **b** $\sigma'_{3c} = 100$ kN/m^2

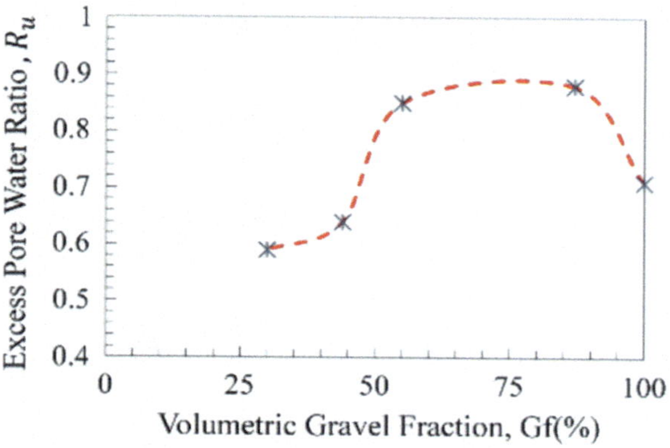

Fig. 3 Maximum excess pore water ratio of GTCM with different Gf

gravel particles primarily form the matrix, the replacement of some gravel particles with softer, less stiff tire chips affects the mixture's properties. The GTCM sample with Gf = 87% behaves like gravel in a relatively loose state, resulting in lower liquefaction resistance compared to the pure gravel sample. The most effective gravel fraction for enhancing liquefaction resistance in the GTCM samples was found to be around 50%. At this proportion, the mixture displayed a significant improvement in liquefaction resistance, indicating an optimal balance between the gravel and tire chip components.

In Fig. 4, the shear modulus curves of GTCM samples with different Gf are presented. For GTCM specimens with a Gf of 100 and 87%, there was a significant drop in shear modulus as shear strain increased over the initial few loading cycles. This decline in shear modulus, correlating with increasing shear strain, is attributed to the diminished contact among gravel particles, caused by the swift accumulation of pore water pressure during cyclic loading. Nonetheless, the rubber content did not markedly influence the shear modulus of samples with the Gf from 55 to 30%, with the shear modulus of these specimens at higher shear strains appearing nearly the same.

It can be concluded that the gravel fraction of 50–60% exhibited a balanced performance in terms of restraining the rise in excess pore water pressure and maintaining sufficient stiffness. This was evidenced by a lower maximum excess pore water pressure ratio and an effective shear modulus at this gravel fraction range compared to other compositions.

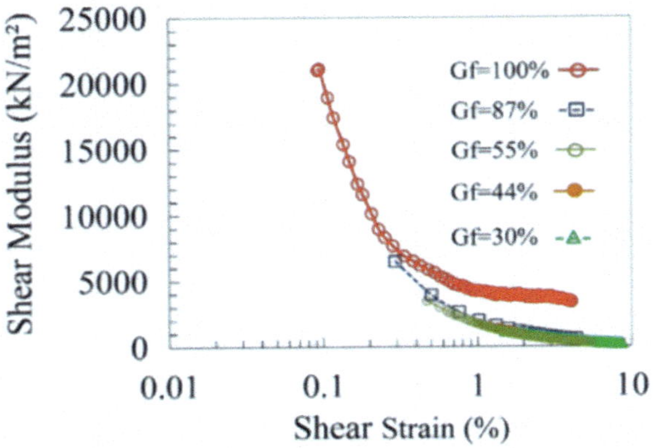

Fig. 4 Effect of gravel fraction on shear modulus reduction curves of GTCM

3 Physical Modeling Tests

3.1 Test Design

The model test was conducted using the 1-*g* shaking table facility at the geo-disaster laboratory of Kyushu University. Soil–structure–fluid interaction was taken into account using the scaling law proposed by Iai [16]. Since this research involves liquefaction-induced damage to on-surface structure, it is the most suitable similitude relationship. In this model test, a geometrical scaling factor of 1:32 was set based on this relationship.

Toyoura sand was used as the foundation soil. A non-liquefiable, dense layer of this sand (relative density D_r = 90%) was created to a depth of 200 mm using dry deposition and tamping methods. Above this, a 300 mm deep liquefiable layer (D_r = 50%) was formed exclusively through the dry deposition technique. The material properties are shown in Table 2. Considering that heavy buildings could settle more from their weight than from liquefaction effects [15], a shallow foundation was simulated using brass rectangular blocks, representing a bearing pressure of 5 kPa and having a model-scale cross-sectional area of 230 mm × 100 mm. For saturation, water was gradually and uniformly percolated through three inlets at the container's bottom.

Table 2 Properties of materials used

Material	Toyoura sand	GTCM
Specific gravity, G_s	2.64	1.91
Dry density, γ_d (g/cm^3)	1.506	1.022
Maximum density, ρ_{max} (g/cm^3)	0.976	1.036
Coefficient of curvature, C_c	0.611	0.842
Coefficient of uniformity, C_u	0.16	4.00

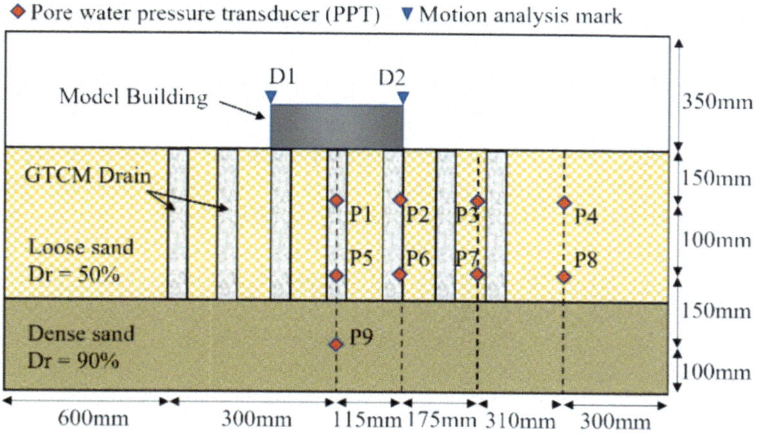

Fig. 5 Cross-sectional view of the model

Two cases were conducted to evaluate the effect of GTCM drains in foundations. Case 1 involved the installation of four parallel prefabricated GTCM drains, each 50 mm in diameter and 300 mm high, placed vertically around the four sides of the residential building model. These drains extended from the surface level to the bottom of the loose sandy layer and further into the denser layer. Cross-sectional view of the tests is depicted in Fig. 5. As contrast, in Case 2, the default condition, no drains were installed.

Both cases were subjected to a sinusoidal acceleration of 200 Gal at a frequency of 4 Hz for a duration of 10 s. Various transducers were used to measure acceleration, pore water pressure, and displacement at different locations, as illustrated in Fig. 5. Pore pressure transducers (PPTs) were strategically positioned to monitor the pore water pressure at specific locations, while motion analysis markers (D1 and D2) were placed at the middle of the short sides on top of the model buildings.

3.2 *Results and Discussion*

Settlement. Significant settlement of the model building in Case 2 was observed, where it was positioned on unreinforced soil. The building experienced considerable sinking, ending up level with the original ground surface. According to motion analysis, the maximum displacements recorded were 40.13 mm at D1 and 32.02 mm at D2, with an observed rotation of about two degrees due to uneven settlement. In stark contrast, the building in Case 1, which incorporated GTCM drains, exhibited markedly less settlement. It remained almost in its original position post-earthquake. The motion marks in this scenario recorded a maximum displacement of just 1.02 mm at D1 and 0.88 mm at D2. This reduced movement suggests that the excess pore water pressures did not reach full liquefaction levels beneath the building in Case 1. The significant difference in settlement between the two cases, nearly 40-fold, underscores the effectiveness of the GTCM drain installation in mitigating earthquake-induced soil liquefaction.

Excess Pore Water Pressure. The settlements of the model buildings, as previously mentioned, were caused by the generation of excess pore water pressures and the resulting liquefaction of the loose sand layer ($D_r = 50\%$). Figure 6 displays the time history of the excess pore water pressure ratio (R_u) measured by PPTs at locations P1, P2, P3, and P4 in two cases. In benchmark case without any drains, only P1 did not reach the full liquefaction threshold of $R_u = 1$. R_u rapidly rose at the onset of the earthquake and remained high throughout the shaking duration. The generation of excess pore water pressure during the earthquake led to liquefaction believed to be the primary cause for the substantial settlement of the structure in this case.

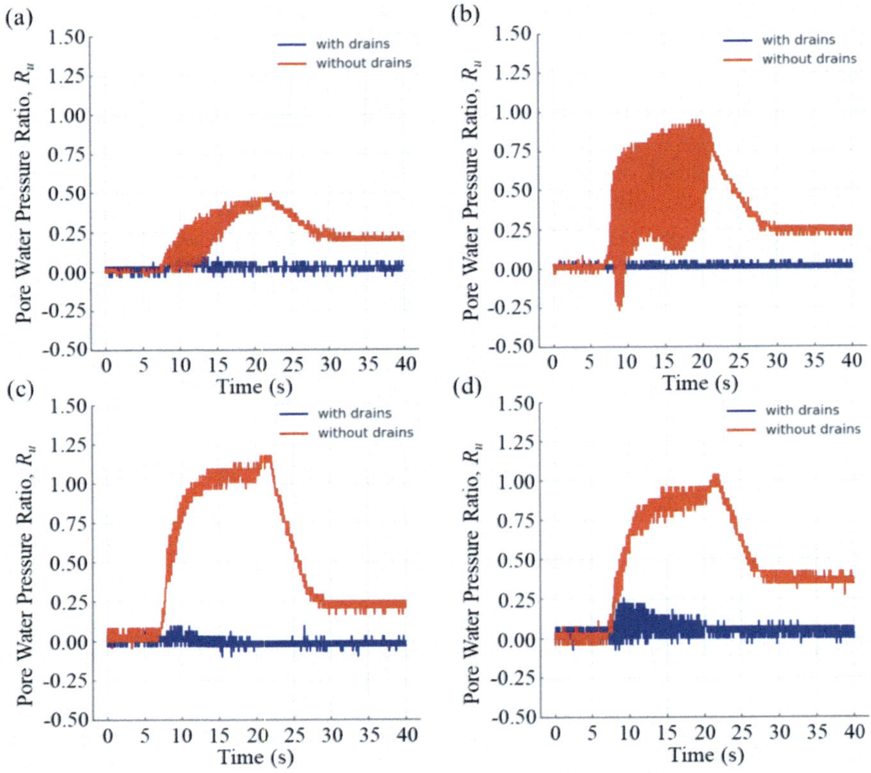

Fig. 6 Comparison of excess pore water pressure ratio at **a** P1; **b** P2; **c** P3; **d** P4

In contrast, Fig. 6 also shows that in Case 1, R_u values remained exceptionally low at all the locations. This indicates that the GTCM drains effectively dissipated excess pore water, significantly contributing to the prevention of liquefaction. This difference between the cases highlights the efficacy of GTCM drains in mitigating liquefaction-related damage during earthquakes.

4 Numerical Method

4.1 *Finite Element Modeling*

Finite element (FE) modeling was carried out in this research using the Swandyne code, which is an effective stress based, fully coupled FE program for solving problems in geotechnical earthquake engineering [2]. The formulation of this code was discussed by Madabhushi and Zeng [23]. The two physical models were enlarged to their prototype scale and simplified into two dimensions in numerical analysis. As shown in Fig. 7, a soil foundation with the depth of 9 m and the length of 26 m was established. A building with a height of 3 m and the length of 8 m was placed on top of the foundation. Two GTCM drains, each with a diameter of 1.5 m, were arranged on either side of the building, with a center-to-center spacing of 3 m between the drains, as a simulation of Case 1.

The FE discretization was carried out using eight nodded isoperimetric, quadrilateral elements with nine gauss points for integration. In total, there were 386 elements in the mesh, 1316 solid nodes, and 430 fluid nodes. As depicted in Fig. 9, the base

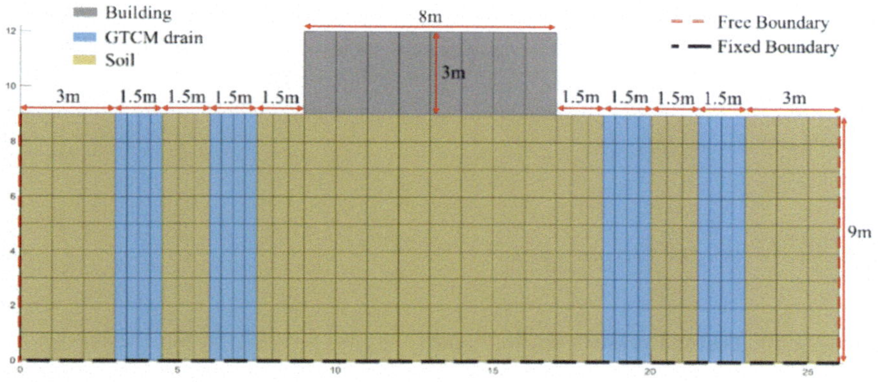

Fig. 7 Dimensions and FE discretization of the model

nodes of the mesh were constrained in both the X and Y directions. This restriction was essential to accommodate the application of earthquake loading as nodal accelerations. As for the edge boundaries, the nodes were modeled as fixed in the X direction and free in the Y direction. The same ground acceleration, as recorded in the physical model tests on 1-g shaking table, was applied.

4.2 Constitutive Models

The constitutive model used for the building and GTCM drains was linear elastic, since failure was not expected. The Young's modulus of the building was taken as 70 GPa. Similarly, for the GTCM drains, the Young's modulus was taken as 250 MPa, since the stiffness of the GTCM is considered to be at the same level as pure gravel, which would be discussed in the following section. As for the Poisson's ratio, the value for sand and drains was taken as 0.15 and 0.3, respectively. The permeability of the GTCM, on the other hand, was set as 0.002 m/s [20].

For modeling soil in FE analyses of geotechnical problems, a key aspect is to take into account the plastic behavior of soil. One of the simplest soil models is a Mohr–Coulomb type model. In the static run, the soil elements were modeled as modified Mohr–Coulomb material to establish the static stresses at the correct g level. However, it does not capture the volumetric strains under cyclic loading, which means no excess pore pressure buildup occurs. Therefore, in order to capture the cyclic behavior of saturated soil, the Pastor-Zienkiewicz Mark III model that was developed by Pastor et al. [26] was used in the dynamic run of the analyses. This model was used by several researchers to conduct liquefaction analysis [4, 21, 24]. The parameters of the materials used are concluded in Table 3. The parameters mentioned have been applied in a broad spectrum of dynamic studies on liquefaction within boundary value problems, as documented by researchers such as Madabhushi and Zeng [23], Haigh et al. [6], Madabhushi et al. [22], and Madabhushi and Madabhushi [25]. This research consistently utilized these established parameters across all analyses without any alterations.

4.3 Results and Discussion

Excess Pore Water Pressure Ratio. Figure 8 shows the distribution of excess pore water pressure ratio at various time steps. This progression provides insight into how the pore water pressure ratio develops and changes over time within the soil. The color gradient represents the value of the excess pore water pressure ratio, with the scale ranging from − 0.5 to 1. The blue color suggests that the majority of the soil domain has low R_u values, indicating that significant pore pressure buildup is not occurring. The results suggest that the vertical drains are efficiently reducing pore water pressure

Table 3 Parameters used for Toyoura sand at a relative density of $D_r = 50\%$

Description	Parameter	Value	Unit
Permeability	k	1×10^{-4}	m/s
Dry density	γ_d	14.50	kN/m^3
Void ratio	e	0.79	–
Critical state friction angle	ϕ'	33	°
Slop of the critical state line	M_g	1.15	–
Yield surface parameter	M_f	0.95	–
Dilatancy parameters	α_g, α_f	0.45	–
Loading plastic modulus	H_0	600	kPa
Unloading plastic modulus	H_{0U}	40,000	kPa
Unloading plastic deformation parameter	γ_{Hu}	2	–
Parameter for reloading discrete memory	γ_{DM}	0	–
Shear hardening parameter 1	p_f	4.2	–
Shear hardening parameter 2	p_g	0.2	–

around them. There does not seem to be any significant buildup of pressure, even around the drains, which would manifest as a higher R_u value in the immediate vicinity of the drains. There is no visible increase in R_u over time, suggesting that there is no ongoing water pressure generation. Since the R_u values are low, the soil stability is likely not compromised by the earthquake loading. The soil structure appears to be stable over the duration of the simulation, with no significant buildup of pore water pressure that would suggest impending failure. Figure 9 suggests there is a significant buildup of excess pore water pressure in the soil, which leads to liquefaction in the absence of GTCM drains improvement measures.

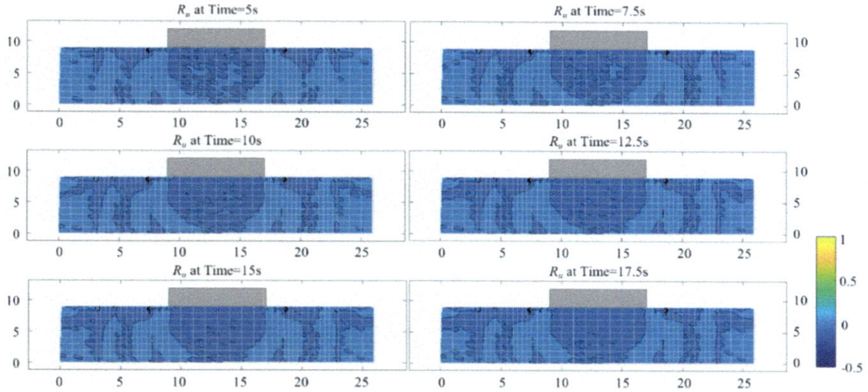

Fig. 8 Excess pore water pressure ratio R_u distribution in Case 1

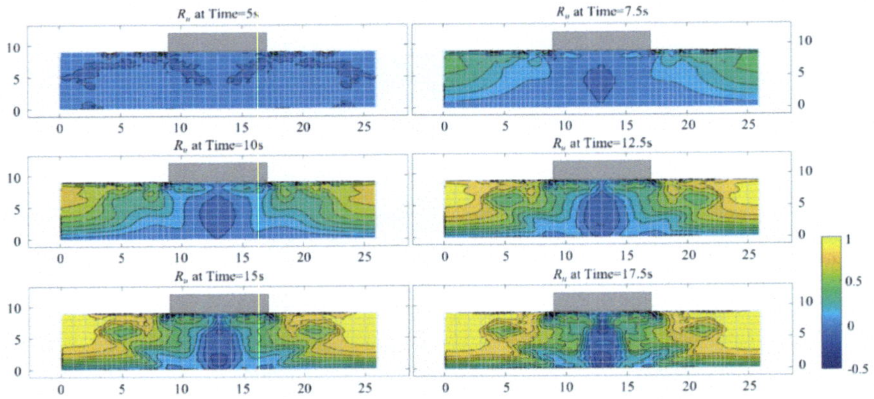

Fig. 9 Excess pore water pressure ratio R_u distribution in Case 2

Despite the absence of liquefaction, the building may still experience some settlement due to the consolidation of the soil as the water drains out. The GTCM drains would be expected to make this process more uniform across the building's footprint, reducing the risk of differential settlement.

Settlement. Figure 10 shows the building settlement over time in both two cases. In Case 1, the building suffered minor settlement during the seismic loading, with improvement of GTCM drains. The maximum settlement of the building at each side was 0.0077 m and 0.0078 m, respectively. The observed settlement patterns in the presence of drains indicate a lower rate of settlement and less variability during the seismic loading. This suggests that the drains are effective in reducing the risk of liquefaction by allowing for quicker dissipation of pore water pressure, thereby maintaining soil strength. In contrast, the absence of drains results in more pronounced settlement changes, particularly during the peak and the winding down phases of the loading. The recorded maximum settlement was 0.0670 m and 0.0696 m on two sides, respectively. The absence of the GTCM drains resulted in the buildup of pore water pressure, reducing soil strength and leading to greater settlement. In summary, the implementation of GTCM drains is shown to be an effective strategy in reducing the risk of earthquake-induced liquefaction. By facilitating the dissipation of excess pore water pressure, the drains help in maintaining soil stability during seismic events.

5 Conclusions

The paper discusses a novel approach to mitigate liquefaction for existing buildings using a gravel tire chips mixture (GTCM) drainage system. The study is grounded in the increasing global interest in using waste tire-derived geomaterials for ground

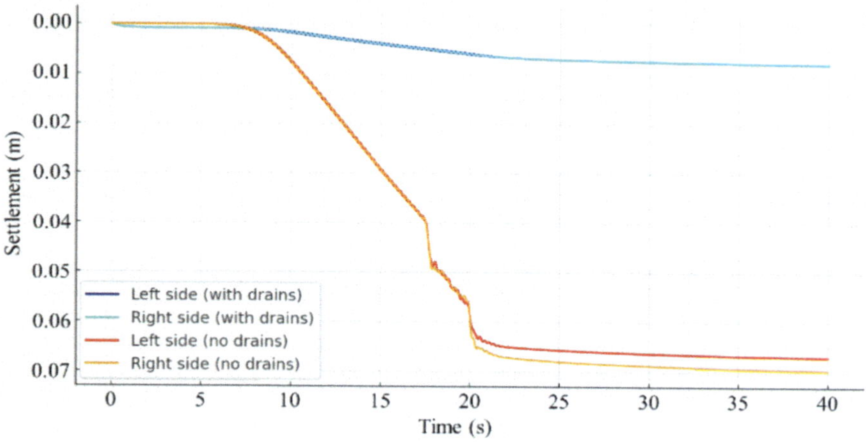

Fig. 10 Comparison of building settlement over time

improvement, with Japan being a leader in this area. The research adopts an integrated approach, combining elements tests, model tests, and numerical simulations to validate the effectiveness of the GTCM drainage system in reducing earthquake-induced soil liquefaction. Key findings of this study are concluded as following:

1. The undrained cyclic behavior of the gravel tire chip mixtures was found to be highly stress dependent. The decrease in the effective confining pressure increases the liquefaction resistance of the GTCM.
2. The gravel fraction plays an important role in the dynamic behavior and liquefaction resistance of GTCMs. 50– 60% is the best mixing percentage, in which the rise in excess pore water pressure can be significantly restrained without compromising the stiffness of the reinforcing inclusion.
3. Significant reduction in settlement and movement of structures during earthquakes when GTCM drains are used, indicating effective mitigation of liquefaction.
4. The GTCM drains effectively dissipated excess pore water pressure, contributing significantly to preventing liquefaction.
5. Numerical simulations further supported these findings, showing stable soil structure and minimal pore water pressure buildup under seismic loading.

Overall, the study presents GTCM as a promising, sustainable technique for liquefaction mitigation, demonstrating its effectiveness through rigorous testing and analysis.

Acknowledgements The authors would like to acknowledge the financial support provided by Kyushu University under Progress 100 project. The work presented in this paper was supported by the JST (Japan Science and Technology Agency) under the "SPRING—Support for Pioneering Research Initiated by the Next Generation" program, (Grant No. JPMJSP2136). The second author would also like to acknowledge the Obayashi Foundation for supporting him in conducting research at the University of Cambridge. Mr. Yuichi Yahiro, technical assistant of Geo-disaster Prevention Laboratory of Kyushu University, is also acknowledged for his help and support while conducting experiments.

References

1. Brennan AJ, Madabhushi GSP (2006) Liquefaction remediation by vertical drains with varying penetration depths. Soil Dyn Earthq Eng 26(5):469–475
2. Chan AHC (1988) A generalised fully coupled effective stress based computer procedure for problems in geomechanics. Swandyne User Manual, Swansea, UK
3. Edil T, Bosscher P (1994) Engineering properties of tire chips and soil mixtures. Geotech Test J 17(4):453–464
4. Español-Espinel C, Haigh SK, Madabhushi GSP (2023) Liquefaction analysis of soil plugs within large diameter monopiles using numerical modelling. Bull Earthquake Eng 21:5443–5458
5. Garcia-Torres S, Madabhushi GSP (2019) Performance of vertical drains in liquefaction mitigation under structures. Bull Earthq Eng 17(11):5849–5866
6. Haigh SK, Ghosh B, Madabhushi GSP (2005) Importance of time step discretisation for nonlinear dynamic finite element analysis. Can Geotech J 42(3):957–963
7. Hazarika H (2007) Structural stability and flexibility during earthquakes using tyres (SAFETY)—a novel application for seismic disaster mitigation. In: Hazarika, Yasuhara (eds) Scrap tire derived geomaterials-opportunities and challenges. Taylor and Francis, London, pp 115–125
8. Hazarika H, Fukumoto Y (2016) Sustainable solution for seawall protection against tsunami-induced damage. Int J Geomechanics 16(5)
9. Hazarika H, Kohama E, Sugano T (2008) Underwater shake table tests on waterfront structures protected with tire chips cushion. J Geotech Environ Eng 134(12):1706–1719
10. Hazarika H, Kuribayashi K, Kuroda S, Hu Y (2023) Performance evaluation of waste tires in protecting embankment against earthquake loading. Bull Earthq Eng 21:4019–4035
11. Hazarika H, Pasha SMK, Ishibashi I, Yoshimoto N, Kinoshita T, Endo S, Karmokar AK, Hitosugi T (2020) Tire chips reinforced foundation as liquefaction countermeasure for residential buildings. Soils Found 60(2):315–326
12. Hazarika H, Yasuhara K (2007) Scrap tire derived geomaterials—opportunities and challenges. Taylor & Francis, London, UK
13. Hazarika H, Yasuhara K, Kikuchi Y, Karmokar AK, Mitarai Y (2010) Multifaceted potentials of tire derived three dimensional geosynthetics in geotechnical application and their evaluation. Geotext Geomembr 28(3):303–315
14. Hu Y, Hazarika H, Madabhushi GSP, Haigh SK (2023) Geosynthetics wrapped tire derived materials as drains for liquefaction mitigation. In: Proceedings of the 12th international conference on geosynthetics (12ICG), Rome, Italy
15. Hu Y, Hazarika H, Pasha SMK, Haigh SK, Madabhushi GSP (2021) Effect of bearing pressure on liquefaction-induced settlement in layered soils. In: Hazarika H, Madabhushi GSP, Yasuhara K, Bergado DT (eds) Advances in sustainable construction and resource management. Lecture notes in civil engineering, vol 144, Springer, Singapore, pp 261–270
16. Iai S (1989) Similitude for shaking table tests on soil-structure-fluid model in 1g gravitation field. Soils Found 29(1):105–118
17. JGS (2009) JGS 0111-2009: test method for density of soil particles, Standards of the Japanese Geotechnical Society. Japanese Geotechnical Society, Tokyo, Japan
18. Karmokar AK (2007) Use of scrap tire derived shredded geomaterials in drainage application. In: Hazarika, Yasuhara (eds) Scrap tire derived geomaterials—opportunities and challenges. Taylor and Francis, London, pp 127–138

19. Karmokar AK, Takeichi H, Yasuhara K (2002) Triaxial compression test on granulated scrap tire rubber for their use as lightweight geomaterials. In: Proceedings of the international workshop on lightweight geo-materials, Tokyo, Japan, pp 127–134
20. Kaushik MK, Kumar A, Bansal AK (2015) Performance assessment of gravel-tire chips mixes as drainage layer materials using real active MSW landfill leachate. Geotech Geol Eng 33:1081–1098
21. Kumari S, Sawant VA (2021) Numerical simulation of liquefaction phenomenon considering infinite boundary. Soil Dyn Earthq Eng 142
22. Madabhushi GSP, Cilingir U, Haigh SK, Hazarika H (2008) Finite element modelling of the seismic behaviour of water front structures. In: The 12th international conference of international association for computer methods and advances in geomechanics (IACMAG), Goa, India
23. Madabhushi GSP, Zeng X (1998) Behaviour of gravity quay walls subjected to earthquake loading. Part II: Numerical modelling. JNL Geotech Eng 124:418–428
24. Madabhushi GSP, Zeng X (2007) Simulating seismic response of cantilever retaining walls. J Geotech Geoenviron Eng 133(5):539–549
25. Madabhushi SSC, Madabhushi SPG (2015) Finite element analysis of floatation of rectangular tunnels following earthquake induced liquefaction. Indian Geotech J 45:233–242
26. Pastor M, Zienkiewicz OC, Leung KH (1985) Simple model for transient soil loading in earthquake analysis. II. Non-associative models for sands. Int J Numer Anal Meth Geomech 9:477–498
27. Yasuhara K (2007) Recent Japanese experiences on scrapped tires for geotechnical applications. In: Hazarika, Yasuhara (eds) Keynote paper, Proceedings of the international workshop on scrap tire derived geomaterials—opportunities and challenges. Taylor & Francis, London, pp 17–40
28. Yasuhara K, Hazarika H, Fukutake K (2006) Applications of scrap tires in civil engineering. Doboku Gijutsu 61(10):79–86 (in Japanese)

Effect of High Groundwater on Flexible Culvert Response to Truck Loading

Ian Moore, Oliver Kearns, and Neil Hoult

Abstract Design of flexible culverts to support vehicle loads is routinely based on soil behaviour where the groundwater table is assumed to be at or below the culvert invert (so that pore pressures are negligible in the backfill beside and above the pipe). To investigate the potential limitations of this assumption, a new facility has been developed at Queen's University which permits high groundwater experiments to be conducted on corrugated steel culverts (where groundwater can be controlled within the test chamber). The effects of high groundwater (like those during a storm) are being studied, where pore pressures reduce the effective stresses and modulus of the soil near the pipe structure, which change the soil–pipe interaction. The ability to introduce and control groundwater is also allowing experiments studying soil erosion beside the test pipe, as a result of inflows through perforations in the pipe walls (due to simulated corrosion). The facility is being used to study pipe responses to vehicle loading (represented by a pair of wheels on a truck axle) under normal groundwater conditions where the water table is at or below the pipe invert, under high groundwater conditions (where the groundwater level is near the ground surface), and after the development of soil erosion. These studies are of importance to highway engineers responsible for assessing culvert capacity and setting load ratings.

Keywords Flood effects · Culvert · Corrugated steel pipe · Soil–pipe interaction · Prototype scale testing

1 Introduction

The design of flexible culverts needs to address the impact of both earth loads and vehicle loads, and calculations of expected values of thrust, and in some cases moment, are undertaken using an assumption that groundwater levels in the backfill are at or below the level of the culvert's invert, Scenario I shown in Fig. 1, and

I. Moore (✉) · O. Kearns · N. Hoult
Queen's University, Kingston, ON K7L 3N6, Canada
e-mail: ian.moore@queensu.ca

B. Indraratna and C. Rujikiatkamjorn (eds.), *Recent Advances and Innovative Developments in Transportation Geotechnics*,
https://doi.org/10.1007/978-981-97-8245-1_15

thus that pore pressures in the soil surrounding the pipe are zero. This influences the magnitude of effective stresses in the backfill in the vicinity of the pipe and the choice of soil modulus in any calculations of soil–pipe interaction (either explicitly or by use of soil modulus tables that were prepared based on this zero-pore pressure assumption, e.g. [1]). During storm events, however, water levels may increase to the level of the crown of the culvert or even at or above the road pavement, Scenario II shown in Fig. 1b, which will reduce the effective stresses in the backfill, and therefore, both the modulus and strength of the coarse-grained backfill. As flood events become more commonplace, highway engineers responsible for load rating of these structures may need to consider how high water tables can impact the thrust and moment demands that result from vehicle loading and the effect of these on the vehicle load capacity of the culvert.

It is well known that corrugated steel culverts corrode over the course of their service life, and the corrosion may lead to perforations through the corrugated steel

Fig. 1 Corrugated steel culvert under truck load, showing two potential groundwater levels in an embankment around a corrugated steel culvert: I—normal operation when the water table is at or below the invert; II—during storm events, when water may rise to near or at the ground surface (I. Moore, used with permission)

a. Perforations along haunches (A. de Oliveira)

b. Erosion beside a culvert due to ingress of groundwater (I. Moore)

Fig. 2 Corroded steel culverts in Ontario (images used with permission)

plates, typically along the haunches of the structure, Fig. 2a. Ingress of groundwater from beside the culvert can then cause erosion of the surrounding soil, with sinkhole development in extreme cases, Fig. 2b. Highway engineers may also consider how local loss of ground support as a result of this erosion can compromise the structure—changing the structural demands (moment and thrust) that develop as a vehicle passes overhead, and jeopardizing the capacity to support truck loading on the ground surface [13, 14].

This paper describes the use of the Deep Burial Simulator at Queen's University, a new test facility developed by Mai et al. [7] to study pipe performance and load capacity at deep burial, and subsequent work reported by Kearns et al. [5] to extend the capabilities of the facility to study pipe behaviour when the ground becomes saturated. The paper introduces the Deep Burial Simulator and then provides an overview of the approach used to control groundwater. It then introduces recent experiments quantifying how corrugated steel pipe response to vehicle loading is influenced by both changes in the effective stresses in the backfill, and soil erosion due to ingress of groundwater.

2 Development of a Test Pit to Examine Buried Pipe Response to High Values of Earth and Vehicle Loading

Mai et al. [7] and Mai [8] describe the conceptual design of the Deep Burial Simulator at Queen's University—a test facility for studying the behaviour of pipes responding to earth loads. The facility has the following features:

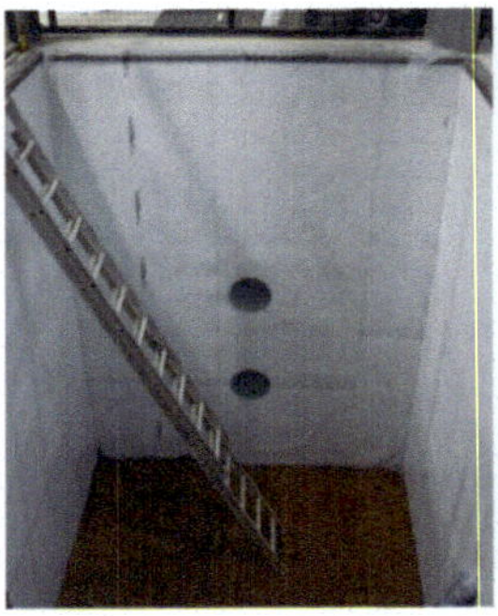

a. Walls covered with UHMWPE (Mai, 2019)

b. Installation of two reinforced concrete pipes (Monteith, 2020)

c. Backfilling steel ribbed polyethylene pipes within the test pit (I. Moore)

d. Surface loading with two 250 t actuators onto steel grillages (I. Moore)

Fig. 3 The 5 m × 3 m × 4.6 m deep burial simulator (images used with permission)

a. The test pit is 5 m long, 3 m wide, and 4.6 m deep, Fig. 3a and b.
b. Sidewalls are coated with sheets of ultra-high molecular weight polyethylene that are commonly used to produce low friction bridge bearings. When further covered with a polyethylene sheet, it reduces the angle of friction on the sidewalls to 11°. When the interface between the two polyethylene layers is lubricated, it reduces the angle of friction to 5°, Fig. 3a.
c. Test pipes can be of 5 m length and up to 1 m diameter (Fig. 3b), or 3 m length and up to 3 m in diameter (Fig. 3c).
d. Two 2500 kN actuators are attached to a reaction frame overhead (Fig. 3d)
e. Two stiff steel grillages which can be placed at the ground surface and loaded by the actuators through bespoke moment-release bearings (Fig. 3d).
f. Overburden pressures equivalent to 22 m of coarse-grained fill can be simulated.

To date, tests have been conducted in the facility on steel-reinforced polyethylene pipes (e.g. [10]), corrugated steel pipes (e.g. [5, 9]), and reinforced concrete pipes

[12]. The facility was designed to have low permeability reinforced concrete sidewalls, so testing can be conducted with saturated ground, and this feature will be explored in this paper.

The original test pit construction included a shaft providing access to the pit through two portholes in the reinforced concrete wall along the northern side, Fig. 3a. Prior to undertaking the controlled groundwater experiments, the lower porthole was enlarged from its original diameter of 360 mm, to become an oval of height 400 mm and width 610 mm. This improved access to any test pipe running from the northern end installed at this level.

3 Control of Groundwater Pressure and Vehicle Load Simulation

Groundwater levels in the test pit are controlled using six plain PVC standpipes installed around the perimeter of the test pit, Fig. 4a, for subsequent use as injection wells—one down each corner of the rectangular pit, and one down the midpoint of each of the two longer sides. Water was added to the pit through those PVC pipes with level controlled using a float valve, Fig. 4b.

Figure 5 presents photographs of the test pit configuration for the corrugated steel pipe experiments described in the subsequent sections. Both images show some of the PVC standpipes.

Figure 5b shows the configuration used to simulate vehicle loading at the ground surface. Vertical load is applied through a steel plate of 600 mm length and 250 mm width, which is the standard dimension of a wheel pair defined in the Canadian Highway Bridge Design Code [2]. This represents the wheel pair at one end of a truck axle. Vertical forces are applied using the actuators shown in Fig. 3d, through a steel column, placed on the steel plate. Locations of the loading plates are included in the plan view of the test pit given in Fig. 4a.

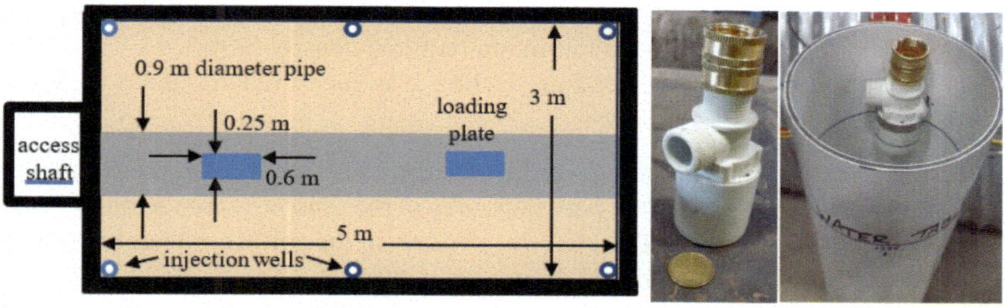

a. Test layout (adapted from Kearns, 2022) b. Float valves (Kearns, 2022)

Fig. 4 Groundwater control system, including injection wells (standpipes) and float valves

a. Pipe placed on bedding **b**. Vertical force on a steel plate to simulate a wheel pair

Fig. 5 Test setup in the deep burial simulator at Queen's University; vertical standpipes are used to control the groundwater level [4]

4 Saturated Ground Experiments

Figures 4a, 5, and 6 show the layout of the buried corrugated pipe experiments undertaken with the ground saturated. A 0.9-m diameter corrugated steel pipe specimen has been examined, fabricated by helical winding of corrugated steel plate with depth of 13 mm, pitch of 68 mm, and wall thickness of 1.82 mm.

An air-dried, uniform sand was used to backfill the test pit. After installing the foundation soil, the corrugated steel culvert was placed and backfilled. The sand was compacted to approximately 97% of the maximum dry unit weight achieved in a standard Proctor test—higher than the 90% minimum specified by AASHTO [1], but bringing the uniform sand to a modulus more like that expected for a graded granular backfill compacted to the minimum required density (90%).

The culvert response to wheel-pair loading was measured for three different groundwater conditions, as follows:

1. Response to wheel-pair loading was first measured for what are the conventional design conditions—where the pore pressures in the sand are zero (actually equal to atmospheric pressure). The effective stresses in the soil were therefore equal to the total stresses—the normal design assumption.
2. The standpipes were used to inject water and raise the water table above the crown to just below the ground surface, and measurements were again taken of culvert response to wheel-pair loading. This is a useful proxy for the groundwater conditions that are expected in a storm or flood event. High stream flows on the

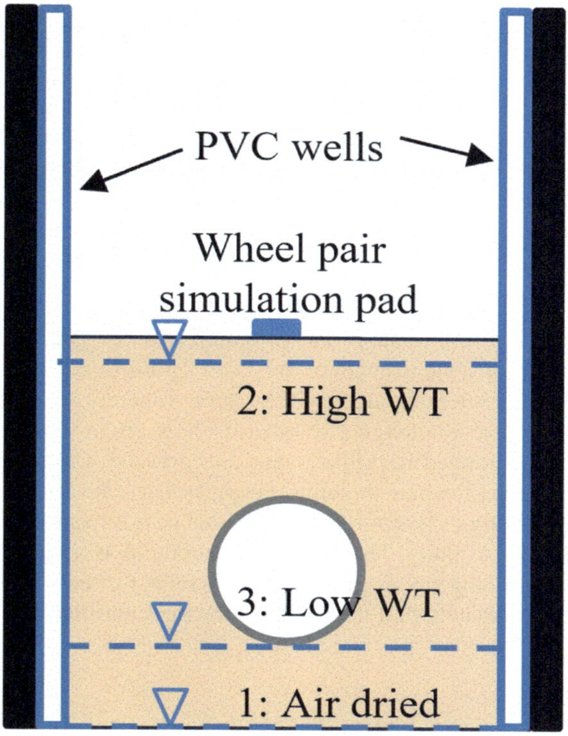

Fig. 6 Layout for the buried corrugated steel pipe test showing the three water table (WT) positions: 1—at the base of the pit (the air-dried condition), 2—near the ground surface, and 3—at the invert

upstream side of the culvert may overwhelm its hydraulic capacity and build up (pond) upstream. Since coarse-grained soils are normally used to backfill corrugated steel pipes and these have relatively high hydraulic conductivities, water may then readily flow from that pond into the embankment and raise the groundwater level within the embankment soil outside the pipe, up towards the ground surface. In some cases, the water level may extend above the upper surface of the pavement. Depending on the time over which the storm or flood persists, and whether there is any risk of culvert washout, trucks may continue to use the creek crossing during these high groundwater level events. The response of the corrugated steel pipe to loading applied when groundwater level is high should be significantly different to that when the soil was dry. Since bulk unit weight of soil is approximately twice the unit weight of water, the high values of pore pressure will approximately halve the effective stresses in the backfill around the pipe. This will reduce both the shear strength and modulus of the soil. The magnitude of elastic soil modulus strongly influences the soil–pipe interaction of buried corrugated steel pipes (see [6, 11]). Reduced effective stress will decrease the shear strength of the soil and increase the risk of bearing failure under the action of wheel loads.

3. The final groundwater condition examined in the experiments was obtained by lowering the water in the standpipes back to the level of the invert of the test pipe. This represents the soil conditions expected after floodwater has receded, and water has drained from the embankment. In most cases, the water table will return to a level near the culvert invert, or below that level. The moisture remaining in the embankment soil has potential to generate suctions (particularly if the backfill contains fine sand or silt particles). Comparisons between the culvert responses to surface loading for this unsaturated soil condition and for the dry soil condition can indicate the extent to which suctions in the test sand influenced soil modulus and strength.

If a conventional Mohr–Coulomb criterion is used for backfill in saturated or dry conditions, the strength of the coarse-grained material will be dominated by friction, and that shear strength will be approximately a linear function of effective stress. It is more challenging to quantify how effective stress influences the modulus of coarse-grained backfills (sands and gravels). Consider however the approximation of Janbu [3] where modulus is expressed as a function of confining stress to the power of n. Selig [15] reports that the value of n for sand at the maximum Proctor density is close to unity. This means that modulus is also approximately a linear function of confining stress, and backfill modulus for the saturated ground condition will be approximately half the value for the dry condition.

5 Erosion Experiments

Kearns et al. [5] describe how the facility is being used to conduct erosion experiments. Following completion of the saturated ground experiment, the corrugated steel pipe was modified to uncover a series of perforations that were prepared along the haunches prior to burial, Fig. 7a. The groundwater level in the test pit was then raised to the crown, and this initiated ingress of groundwater and the sand backfill through the perforations, Fig. 7b. Erosion was then stopped by lowering the water table back to the invert, and the corrugated steel culvert responses to surface load were measured for this eroded soil condition. The volume of eroded sand was monitored during the experiment by directing the eroded material from the end of the pipe into a barrel located within the access shaft, and occasionally removing sand that had accumulated within the culvert.

Further testing to explore other erosion void geometries was undertaken by raising and subsequently lowering the water table. This included work to measure response when the eroded soil volume approximately doubled. It also included work to study the impact of erosion from perforations along the haunch on one side of the corrugated steel pipe and erosion from perforations along both sides. Measurements of culvert response for the different erosion geometries were then compared to those made when the ground was intact.

a. Perforations b. Far half of perforated culvert has eroded soil collecting

Fig. 7 Perforated pipe before and during erosion experiment [4]

Following completion of the load testing, the voids were filled with polymer foam so that the void geometries could be recorded after exhumation, Fig. 8. Kearns et al. [5] provide further details of the testing procedures used to study the effect of erosion into the perforated corrugated steel culvert and a reinforced concrete pipe featuring a leaking joint (the pipes shown in Fig. 3b).

6 Conclusions

A testing facility has been developed to explore the effects of groundwater on the response of buried pipes. Experiments are being conducted to investigate how groundwater influences the soil–pipe interaction of a corrugated steel pipe buried in coarse-grained backfill. Three different groundwater conditions are being modelled to investigate key conditions that can occur in the field.

1. When the soil is almost dry, directly after it was placed and compacted.
2. When groundwater table is raised towards the surface to saturate the soil.
3. When the groundwater subsequently drops back to the level of culvert invert.

The modulus and shear strength of the soil decrease as effective stresses reduce, and this influences how soil–pipe interaction develops. These experiments can therefore quantify how groundwater conditions influence the increments in bending

Fig. 8 Corrugated steel pipe after exhumation, showing the geometry of the erosion voids which were filled with foam prior to excavation (Moore, used with permission)

moment, thrust, and deformation that develop due to the application of loads simulating the action of a wheel pair on the ground surface over the pipe (vertical force applied to a steel plate). The observed behaviour can guide recommendations regarding the management of load ratings during high groundwater events (e.g. the implementation of truck weight limits during and for some time after storms or floods). Experiments are also being conducted to examine the impact of soil erosion (e.g. through perforations along the haunches of a corrugated steel pipe). Erosion of sand backfill will initiate once groundwater is raised above the crown of the structure. Culvert responses to simulated vehicle loading can then be compared for the intact soil condition (before there is any erosion), and after different amounts of erosion.

Work is ongoing to understand the implications of pipe responses (peak bending moments and thrusts under surface loading) after backfill erosion and to assist with the development of load rating guidelines for use by highway engineers responsible for these kinds of structures. The test facilities can also be used to study other types of pipe (e.g. clay, concrete, and polyethylene sewers with leaking joints) and to evaluate the effect of soil erosion on the performance of pipes at deep cover.

References

1. AASHTO (American Association of State Highway and Transportation Officials) (2017) Section 12: Buried structures and tunnel liners. AASHTO LFRD Bridge Design Specifications, LRFD. AASHTO, Washington, DC
2. CSA (Canadian Standards Association) (2017) Canadian highway bridge design code. CSA, Mississauga, Ontario, Canada
3. Janbu N (1963) Soil compressibility as determined by oedometer and triaxial test. In: Proceedings of the European conference on soil mechanics and foundation engineering, pp 19–25
4. Kearns O (2022) Effects of soil cover, saturation, and erosion voids on the live load response of corrugated steel culverts. Ph.D. thesis, Civil Engineering, Queen's University, 217 p
5. Kearns O, Moore ID, Hoult NA, Monteith T (2024) Development of testing procedures to study the impact of erosion voids on pipes. Geotech Test J
6. Liu Y, Moore ID, Hoult NA (2024) New design equations for corrugated steel culverts responding to live loads. Can Geotech J 61(1):1–15
7. Mai TV, Hoult NA, Moore ID (2018) Design and performance of a deeply buried pipe testing facility. Adv Struct Eng 21(16):2571–2588
8. Mai TV (2019) The development and use of a test facility for large diameter pipes. Ph.D. Thesis, Department of Civil Engineering, Queen's University, Kingston, Ontario, Canada
9. Mai VT, Hoult NA, Moore ID (2021) Laboratory investigation of the structural performance of a corrugated steel culvert under increasing cover depth. J Bridge Eng 26(6), 04021029:1–11
10. Mai VT, Hoult NA, Moore ID (2021) Laboratory investigation of the structural performance of a large diameter steel reinforced polyethylene pipe. J Pipeline Syst Eng Manage 15(1)
11. Moore ID (2001) Culverts and buried pipelines. In: Rowe RK (ed) Geotechnical and geoenvironmental handbook. Kluwer Acad. Publishers, pp 541–568
12. Monteith TR (2020) Development and influence of erosion voids at rigid pipe joints. MASc Thesis, Department of Civil Engineering, Queen's University, Kingston, ON, Canada
13. Peter J, Chapman DN, Moore ID, Hoult NA (2018) Impact of soil erosion voids on reinforced concrete pipe responses to surface loads. Tunnel Undergr Space Technol 118:111–124
14. Peter J, Moore ID (2019) Effects of an erosion void on a deteriorated metal culvert before and after repair with a grouted slip liner. J Pipe Syst Eng Manag 10(4)
15. Selig ET (1998) Soil parameters for design of buried pipelines. In: Pipeline infrastructure—proceedings of the conference. American Society of Civil Engineering, New York, NY, pp 99–116

Recent Advances in Computational Methods, Predictions, and Applications to Critical Zones in Railways

Sanjay Nimbalkar, Piyush Punetha, Naveen Kumar Meena, and Mohammad Adnan Farooq

Abstract Accurate evaluation of the mechanical response of railway tracks subjected to train-induced repeated loading is a prerequisite for their proper design and maintenance planning. A multitude of approaches have been developed over the years to accurately predict some of the most crucial aspects of track behaviour. This theme paper presents the recent advances in computational methods that can be employed to predict the performance of railway tracks and demonstrates their capabilities. Firstly, the suitability of the finite element (FE) method in understanding the soil arching mechanism, which governs the behaviour of the pile-supported railway embankments, is demonstrated. The influence of pile spacing on the soil arching mechanism is illustrated. Subsequently, the utilisation of the FE method in evaluating the performance of the ballasted and slab tracks is explored. The vertical stress variation and the accumulation of settlement in both tracks are compared. In addition, a novel computational approach based on a geotechnical rheological track model is presented that can be employed to predict the long-term performance of a standard section of the railway track as well as the critical zones, such as bridge-embankment transitions. Finally, the practical applications of the computational techniques and their future scope are discussed.

Keywords Computational methods · Railway tracks · Critical zones

S. Nimbalkar (✉) · P. Punetha · M. A. Farooq
School of Civil and Environmental Engineering, University of Technology Sydney, Ultimo, Australia
e-mail: Sanjay.Nimbalkar@uts.edu.au

N. K. Meena
Beca Pty. Ltd., Sydney, Australia

B. Indraratna and C. Rujikiatkamjorn (eds.), *Recent Advances and Innovative Developments in Transportation Geotechnics*,
https://doi.org/10.1007/978-981-97-8245-1_16

Notations

$\boldsymbol{C}$, $\boldsymbol{K}$, and $\boldsymbol{M}$	Damping coefficient, stiffness, and mass matrices, respectively
d	Pile width
$d\boldsymbol{w}$, $d\dot{\boldsymbol{w}}$, and $d\ddot{\boldsymbol{w}}$	Displacement, velocity, and acceleration increment vectors, respectively
$\boldsymbol{F}$	Force vector
h	Embankment height
h_{b}	Ballast thickness
h_{n}	Normalised embankment height
N	Number of load cycles
S_{c}	Cumulative settlement in ballasted and slab tracks
s	Pile spacing
s_{e}	Settlement in embankment fill
s_{n}	Normalised settlement in embankment fill
s_{t}	Settlement of track substructure
x	Distance along the track
x_{b}	Distance from the centre of the embankment base
z	Depth below slab/sleeper top
γ	Unit weight of embankment fill
$\delta_{\mathrm{v}}^{\mathrm{e}}$	Vertical elastic displacement in ballasted and slab tracks
σ_{n}	Normalised vertical stress
σ_{v}	Vertical stress

1 Introduction

The railway system plays a crucial role in carrying passengers and transporting resources and goods along major corridors in Australia. To keep up with the growing demand for freight and public transport, the railway industry continues to embrace modern technologies. This has led to faster passenger trains and greater freight train capacity. Consequently, the magnitude and frequency of the train-induced loads on existing railway tracks have significantly increased over the years [1]. Railway tracks have traditionally been perceived as elastic structures comprising superstructure components such as rails and sleepers, which are supported by substructure components like ballast, subballast, and subgrade. However, the increased loads often result in a substantial amount of deterioration and plastic deformation in track substructure layers. This, in turn, may compromise the track geometry and stability, thereby requiring frequent track maintenance [2]. Critical zones, also referred to as transition zones, represent the discontinuities along the railway track where there is an abrupt change in the support conditions. The excessive deformation of the track substructure at these zones results in a rapid degradation of the track geometry.

In order to achieve optimal performance of railway tracks at normal and critical zones, it is essential to develop a design approach that is more reliable, practical, and adaptable. Developing such a design technique necessitates a thorough examination of the mechanical response of track substructure under train-induced repeated loading. Therefore, several researchers have conducted field investigations to comprehend the behaviour of railway tracks and better understand the issues related to track instability and geometry degradation [e.g., 3, 4]. Although field investigations offer valuable insights into the behaviour of railway tracks subjected to repetitive loads, data collection in the site is often a time-intensive and costly process. Laboratory investigation is an alternative approach to field study that involves simulating real-world problems in a controlled environment to recognise the influencing parameters and underlying mechanisms. However, these experiments are generally restricted to small scale due to financial constraints and limited testing space and facilities.

Numerical and analytical methods serve as cost-effective alternatives for understanding the mechanical behaviour of the track substructure. They play a vital role in assessing and improving the performance, safety, and cost-effectiveness of railway tracks through accurate modelling of geometry, material behaviour, and complex boundary conditions. This paper delves into recent applications of advanced computational methods to gain insights into the intricate aspects related to railway track behaviour. Firstly, the suitability of the finite element (FE) method in comprehending the soil arching mechanism, which governs the behaviour of pile-supported railway embankments, is explored. Subsequently, the use of the FE method for comparing the response of ballasted and ballastless (or concrete slab) tracks under train-induced repeated (cyclic) loading is demonstrated. Finally, a novel rheological approach is presented that can predict the long-term performance of a railway track along normal as well as critical zones subjected to moving train loads.

2 Evaluating the Behaviour of Railway Tracks Using FE Method

The FE method is a powerful numerical analysis technique that can be utilised to study the mechanical behaviour of railway tracks under train-induced loading and gain insights into various aspects that influence track response. Consequently, several researchers have attempted to predict the track response using two-dimensional (2D), two-and-a-half dimensional (2.5D), and three-dimensional (3D) FE analyses [5–10]. In this section, the capability of the FE method in addressing some of the most challenging issues pertaining to railway tracks is demonstrated. These include: (a) exploring the soil arching mechanism in railway embankments supported by pile foundations and (b) comparing the performance of ballasted and ballastless (slab) tracks.

2.1 Insights into the Load Transfer Mechanism in Railway Embankments Supported by Pile Foundations

The FE method can be employed to understand the load transfer mechanism in railway embankments supported by piles. Such embankments are commonly built in areas with soft soil that has limited bearing capacity and high compressibility. In these embankments, the settlement is usually greater in the zones between piles compared to that above the piles. This is because there is a significant disparity in the stiffness of piles and surrounding soft soil. However, the settlement is constrained by the mobilisation of shear resistance, which alleviates stress on the soft soil and transfers the majority of the load to the pile foundations. This load transfer mechanism is commonly termed soil arching [11]. The physical visualisation of soil arching is a challenging and resource-intensive process, often necessitating extensive test setups. However, the FE method offers an efficient alternative, providing reasonably accurate results while minimising both time and cost.

To demonstrate the capability of the FE method in gaining insights into the soil arching phenomenon, an equivalent 2D FE model of a pile-supported embankment for railways was developed in ABAQUS [12]. Typically, the analysis of pile-supported railway embankment is a 3D problem, and its transformation to a 2D problem is facilitated via the adoption of an equivalent area method. The axle load is considered as 35 t, which represents loading from a heavy-haul freight train in Australia. The model comprised an embankment fill overlying a gravel bed, which was supported by pile foundation installed in the subsoil (see Fig. 1a). The behaviour of embankment fill, gravel bed, and subsoil was simulated using a linear-elastic, perfectly plastic model with the Mohr–Coulomb yield criterion, whereas the behaviour of the pile was considered linear-elastic. The soil–pile interaction was simulated by employing the Coulomb friction model available in ABAQUS [12]. More details on the methodology can be found in [13].

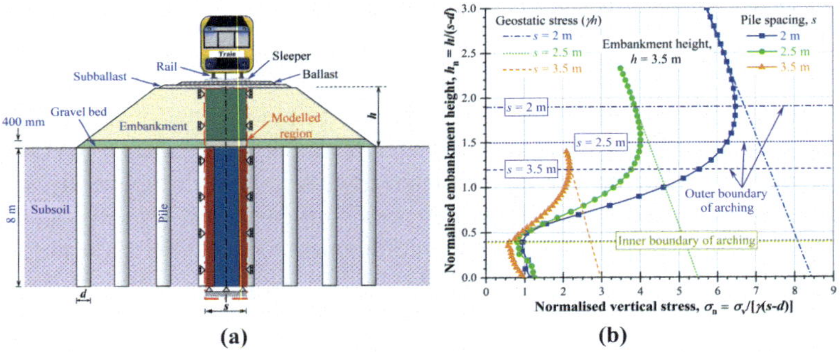

Fig. 1 **a** Railway embankment supported by piles; **b** normalised vertical stress in embankment fill for varied pile spacing (data sourced from [14])

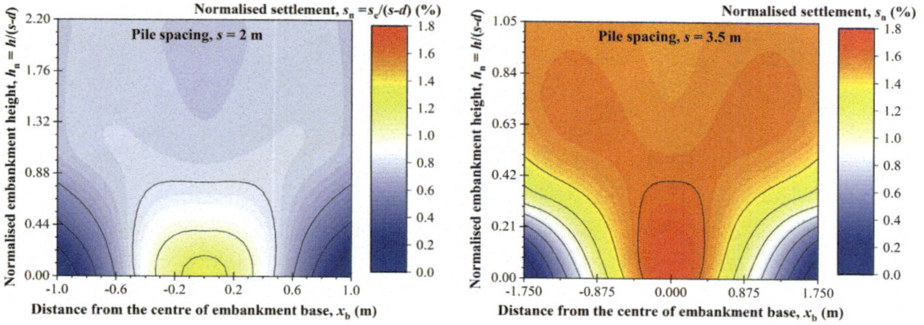

Fig. 2 Normalised settlement contours in embankment fill for varied pile spacings (data sourced from [13])

Figure 1b depicts the variation of normalised vertical stress (σ_n) predicted using the FE method with normalised embankment height (h_n) for three different pile spacings ($s = 2, 2.5$, and 3.5 m). It is apparent from the figure that σ_n initially increases with a slope equivalent to the overburden pressure, spanning from the top of the embankment fill to the outer boundary of soil arching. The normalised vertical stress profile is subsequently disrupted, with a decrease at the middle of the embankment base and an increase at the top of the pile head. Most of the embankment load is effectively transferred to the piles at this stage. It is also evident from Fig. 1b that the inner boundary of the arch remains consistent across all pile spacings. However, the outer boundary exhibits variations corresponding to the pile spacing. The soil arching zone typically enlarges as the pile spacing decreases.

Figure 2 shows the normalised settlement (s_n) contours in the embankment fill predicted using the FE method for $s = 2$ and 3.5 m. It is apparent from the figure that a rise in pile spacing leads to an increase in s_n. For $s = 2$ m, an equal settlement plane (i.e., a horizontal section with no differential settlement) is evident at $h_n = 1.34$; however, no such plane is observed for $s = 3.5$ m due to the absence of soil arching. Thus, the findings from the FE analyses suggest that it is imperative to establish an optimal pile spacing for achieving efficient load distribution in the railway embankments supported by pile foundations.

2.2 *Comparison of the Performance of Ballasted and Slab Tracks*

This section highlights the application of the FE method in comparing the performance of ballasted and concrete slab tracks. To this end, lab-sized 3D FE models for ballasted and slab tracks were developed using ABAQUS [12] (see Fig. 3). In these models, the rail, sleepers, concrete slab, cement asphalt mortar (CAM), and base layer were considered as elastic materials, whereas the ballast, subballast, and

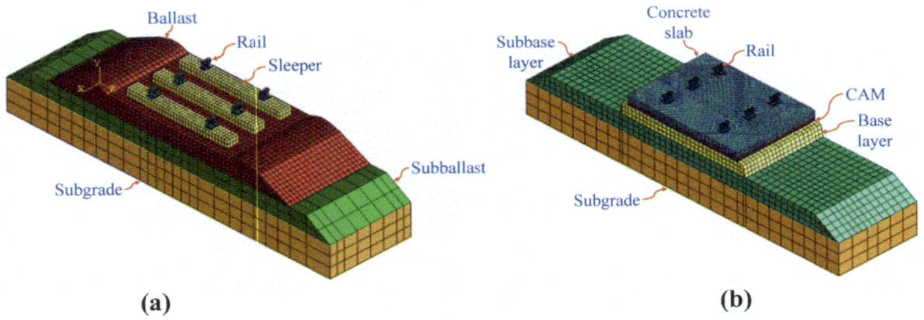

Fig. 3 3D FE models of **a** ballasted and **b** slab tracks

subbase were modelled as elastoplastic materials with the Drucker–Prager yield criterion. The subgrade was represented as a linear-elastic, perfectly plastic material, following the Mohr–Coulomb yield criterion. The interaction between different track layers was simulated using a surface-to-surface contact approach. A detailed explanation of the methodology can be found in previous works [e.g., 15, 23].

Figure 4a shows the distribution of vertical stress (σ_v) and vertical elastic displacement (δ_v^e) with depth in ballasted and slab tracks. The results are provided for the data points located at the interfaces between the different track components, and these are represented as *B* and *S*, followed by the number (which increases with depth) for ballasted and slab tracks, respectively. It is evident from the figure that the peak stress induced by the train load (axle load = 12 t, which represents loading from a Japanese Class-300 high-speed passenger train) on the upper surfaces of both tracks is 5.4 MPa. In the ballasted track, the reduction in σ_v with depth is relatively uniform, with a 74% and 98% decrease from B1 to B2 and B3, respectively. The reduction in σ_v is minimal beneath the ballast layer. In contrast, in the slab track, σ_v decreases by 93.3% and 96.2% from S1 to S2 and S3, respectively. Thus, it is evident that the stress dissipates significantly at a depth of 0.2 m from the slab top in the slab track, in contrast to the ballasted track where a similar stress state exists at a depth of 0.55 m from the top of the sleeper. This difference is attributed to the higher stiffness of reinforced concrete slabs, promoting better stress dispersion. The stress at the bottom of the subgrade in the case of ballasted track (100 kPa) is twice of that in the case of slab track (50 kPa).

It can also be observed from Fig. 4a that δ_v^e at B1 is 1.07 mm in the ballasted track, and it reduces by 46.7% and 98.9% from B1 to B2 and B3, respectively. On the other hand, in the slab track, δ_v^e is 0.058 mm at S1, which reduces by 13.8% and 51.7% at S2 and S3, respectively. It is apparent from the figure that δ_v^e at the top of the slab track is 95% lower than that at the top of the ballasted track.

Figure 4b illustrates the cumulative settlement of both track types with load cycles. It is evident that at the completion of 1.2 million load cycles, the slab track settlement (0.295 mm) is roughly 26 times lower than the ballasted track (7.86 mm). Thus, the results reveal that the slab tracks perform significantly better than the ballasted tracks for the same input loading. However, the selection of the most adequate track type also depends on other aspects, such as cost and environmental impacts.

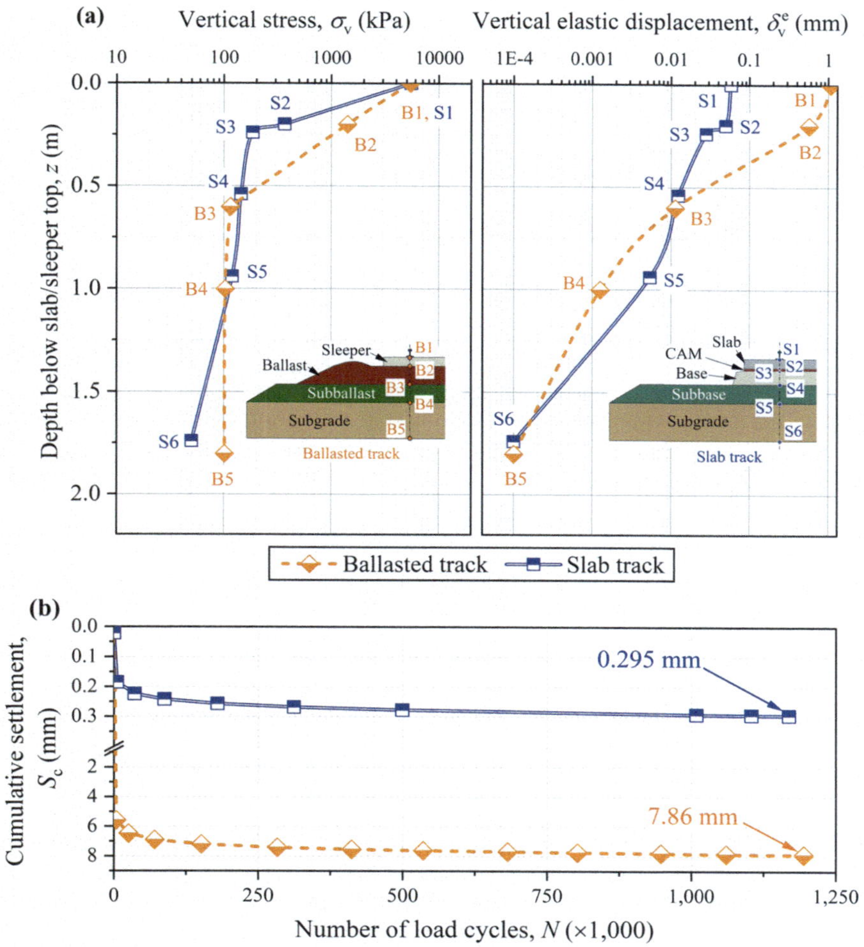

Fig. 4 Comparison of **a** vertical stress and vertical elastic displacement distribution; **b** cumulative settlement for ballasted and slab tracks (data sourced from [15])

3 Geotechnical Rheological Model for Ballasted Railway Tracks

The analytical techniques provide a comparatively quicker and computationally more efficient option than the FE method for accurately predicting the track response, especially when dealing with millions of load repetitions. Consequently, numerous analytical models of varying complexities have been created to evaluate the performance of railway tracks subjected to train-induced repetitive loading [16]. In these models, the geotechnical track layers are represented using equivalent springs, spring-dampers (or dashpots), homogenous or multilayered half-space, or combinations of these elements. While modelling the substructure layers using equivalent springs can predict the total response, it overlooks the interactions among these layers. To address this concern, some researchers modelled the substructure layers as lumped masses connected through viscoelastic elements, such as springs and dampers [17, 18]. However, most of these studies have focused on examining the short-term behaviour of the tracks using conventional elastic track structure approximation, as mentioned in the introduction section of this paper, and overlooked the long-term performance, particularly the settlement that accumulates in the track substructure layers after numerous load repetitions.

Punetha et al. [19] developed a novel rheological approach for predicting both short and long-term performance of the railway track under moving train-induced repeated loading. In this technique, the plastic slider elements were used to simulate the irrecoverable response of the geotechnical layers of the track, while the viscoelastic response was captured using springs and dampers. Thus, the entire railway track was modelled as an array of discrete masses coupled using elements including springs, dampers, and plastic sliders (see Fig. 5). The total response of the substructure layers under train-induced repetitive loading was predicted by solving the following dynamic equilibrium equation [20]:

$$\begin{aligned} \boldsymbol{M} d\ddot{\boldsymbol{w}}_{\mathrm{m}} + \boldsymbol{C} d\dot{\boldsymbol{w}}_{\mathrm{m}} + \boldsymbol{K} d\boldsymbol{w}_{\mathrm{m}} - \boldsymbol{C}^{p} d\dot{\boldsymbol{w}}^{p}_{\mathrm{m}} - \boldsymbol{K}^{p} d\boldsymbol{w}^{p}_{\mathrm{m}} \\ - \boldsymbol{C}'\{d\dot{\boldsymbol{w}}_{\mathrm{m}-1} + d\dot{\boldsymbol{w}}_{\mathrm{m}+1}\} - \boldsymbol{K}'\{d\boldsymbol{w}_{\mathrm{m}-1} + d\boldsymbol{w}_{\mathrm{m}+1}\} \\ + \boldsymbol{C}^{p'}\left\{d\dot{\boldsymbol{w}}^{p}_{\mathrm{m}-1} + d\dot{\boldsymbol{w}}^{p}_{\mathrm{m}+1}\right\} + \boldsymbol{K}^{p'}\left\{d\boldsymbol{w}^{p}_{\mathrm{m}-1} + d\boldsymbol{w}^{p}_{\mathrm{m}+1}\right\} = d\boldsymbol{F} \end{aligned} \tag{1}$$

where $\boldsymbol{F}$ is the force vector; $\boldsymbol{C}$, $\boldsymbol{K}$, and $\boldsymbol{M}$ are the damping coefficient, stiffness, and mass matrices, respectively; $d\boldsymbol{w}$, $d\dot{\boldsymbol{w}}$, and $d\ddot{\boldsymbol{w}}$ denote the vertical displacement, velocity, and acceleration increment vectors, respectively, of the substructure layers; subscript m represents the mth sleeper; and superscript p denotes the plastic component. This approach was subsequently extended to incorporate the inhomogeneous support conditions (typically encountered in the transition zones) [21], principal stress rotation (PSR) effects [22] and effect of geosynthetic reinforcement [20] on track response.

Figure 6 illustrates an example of the track response predicted using the geotechnical rheological model for an embankment-bridge transition considering the effect of PSR. The stiffer side of the track in this example comprised a ballast layer supported on the bridge deck, while the softer side comprised ballast, subballast, and subgrade layers [22]. The axle load is considered as 25 t, which represents loading from a coal freight train in Australia with 100 t wagons. The results are presented for a total tonnage of 20 million gross tonnes (MGT).

It is apparent from Fig. 6 that an increase in the thickness of the ballast layer decreases the differential settlement in the transition zone. The differential settlement between softer and stiffer sides reduces by 85.9% with a rise in ballast thickness from 300 to 750 mm. This decrease is attributed to the wider load distribution facilitated by

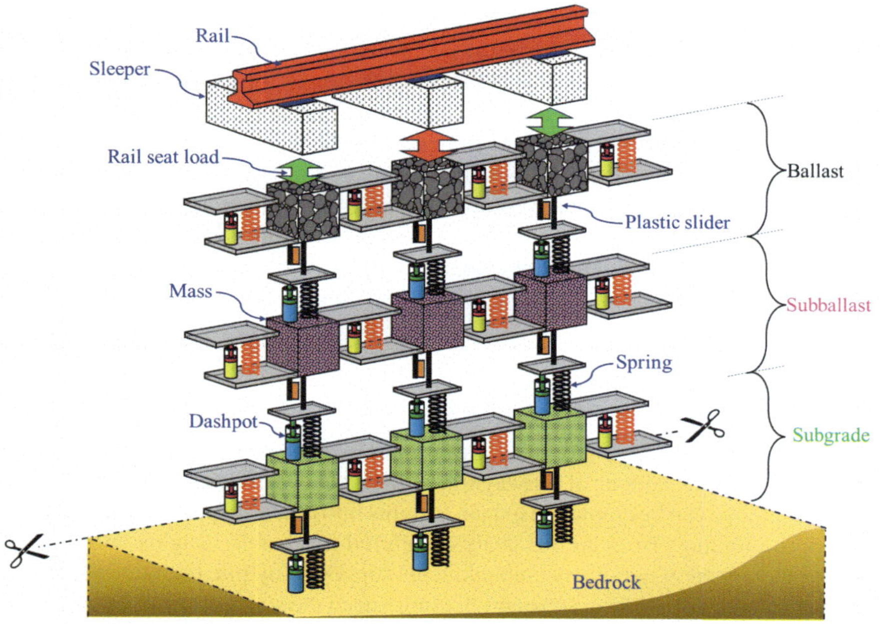

Fig. 5 3D geotechnical rheological track model

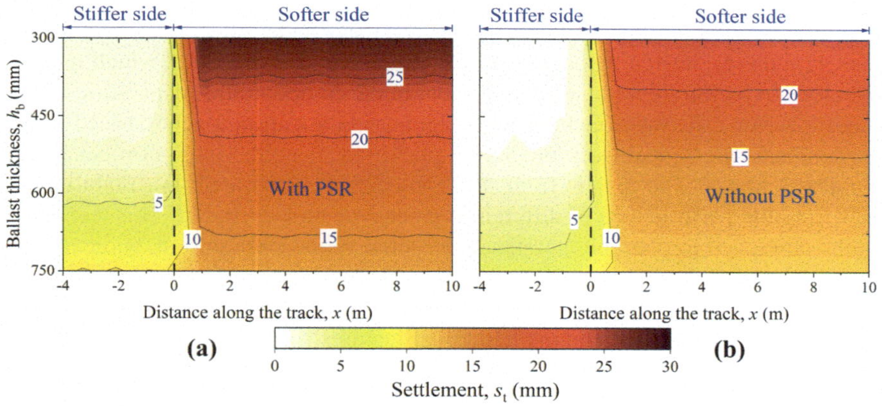

Fig. 6 Effect of ballast thickness on the behaviour of embankment-bridge transition: **a** considering PSR; **b** without PSR (data sourced from [22])

a thicker ballast layer, resulting in the transfer of reduced stresses to the underlying layers. This stress curtailment in the lower layers leads to a reduction in the accumulated settlement. It is also apparent that the differential settlement is considerably greater if PSR is considered in the analysis, in contrast to the case when it is overlooked. These findings underscore the importance of incorporating the PSR effect in the computational approaches to accurately evaluate the response of railway tracks subjected to moving train loads. Additionally, these results demonstrate the practical utility of this rheological approach in examining the suitability of different strategies (e.g., increased ballast thickness) in enhancing the performance of transition zones.

4 Concluding Remarks

This theme paper explored the latest advancements in the computational techniques for analysing the behaviour of railway tracks. Particularly, the paper illustrated how the FE technique can be employed to gain insights into crucial aspects of track performance. The findings from the FE analyses offered invaluable insights into the soil arching phenomenon in railway embankments supported by pile foundations. It was revealed that the pile spacing substantially influenced the soil arching mechanism. In addition, the outcomes of the FE analyses highlighted the significant differences in the cumulative settlement, stress, and elastic displacement variations in the ballasted and concrete slab tracks. Nevertheless, the extensive time required to obtain the long-term response of the railway tracks, particularly under millions of load repetitions, remains an intriguing challenge for the FE-based approaches, which may be addressed in the future with an advancement in the computational capabilities. This issue, however, can also be tackled by employing analytical approaches, based on the geotechnical rheological track model, to quickly evaluate the long-term behaviour of railway tracks. The results from this approach demonstrated its suitability for predicting the performance of different segments of railway tracks. This technique can be employed to plan the maintenance cycles, refine track design, and examine the effectiveness of different approaches in enhancing track performance. Nonetheless, the rheological approach is a relatively new method, and its accuracy is being increased further by incorporating additional features, such as the effect of particle breakage, moisture fluctuations, and strain dependency of elastic modulus on the track behaviour.

References

1. Nimbalkar S, Punetha P, Kaewunruen S (2020) Performance improvement of ballasted railway tracks using geocells: present state of the art. In: Sitharam TG, Hegde A, Kolathayar S (eds) Geocells. Springer Transactions in Civil and Environmental Engineering, Springer, Singapore

2. Li D, Hyslip J, Sussmann T, Chrismer S (2016) Railway geotechnics. Taylor and Francis, Boca Raton, USA
3. Yoo TS, Selig ET (1979) Field observations of ballast and subgrade deformations in track. Transp Res Rec 733:6–12
4. Priest JA, Powrie W, Yang L, Grabe PJ, Clayton CRI (2010) Measurements of transient ground movements below a ballasted railway line. Géotechnique 60(9):667–677
5. Yang LA, Powrie W, Priest JA (2009) Dynamic stress analysis of a ballasted railway track bed during train passage. J Geotech Geoenviron Eng 135(5):680–689
6. Yang YB, Hung HH (2001) A 2.5D finite/infinite element approach for modelling visco-elastic bodies subjected to moving loads. Int J Numer Methods Eng 51(11):1317–1336
7. Li L, Nimbalkar S, Zhong R (2018) Finite element model of ballasted railway with infinite boundaries considering effects of moving train loads and Rayleigh waves. Soil Dyn Earthq Eng 114:147–153
8. Sakhare A, Punetha P, Meena NK, Nimbalkar S, Dodagoudar GR (2023) Dynamic behaviour of integral abutment bridge transition under moving train loads. Transp Geotech 40:100989
9. Punetha P, Maharjan K, Nimbalkar S (2021) Finite element modeling of the dynamic response of critical zones in a ballasted railway track. Front Built Environ 7:1–11
10. Punetha P, Nimbalkar S (2023) Numerical investigation on dynamic behaviour of critical zones in railway tracks under moving train loads. Transp Geotech 41:101009
11. Terzaghi K (1943) Theoretical soil mechanics. Wiley, New York
12. Dassault Systèmes: Abaqus (Version 2018). Dassault Systèmes Simulia Corp, Providence, RI (2018)
13. Meena NK, Nimbalkar S, Fatahi B, Yang G (2020) Effects of soil arching on behavior of pile-supported railway embankment: 2D FEM approach. Comput Geotech 123:103601
14. Nimbalkar S, Meena NK (2022) Static and seismic assessment of soil arching in piled embankments. In: Kolathayar S et al (eds) Civil engineering for disaster risk reduction. Springer Nature, Singapore, pp 263–281
15. Farooq MA, Nimbalkar S, Fatahi B (2021) Three-dimensional finite element analyses of tyre derived aggregates in ballasted and ballastless tracks. Comput Geotech 136:104220
16. Knothe KL, Grassie SL (1993) Modelling of railway track and vehicle/track interaction at high frequencies. Veh Syst Dyn 22(3–4):209–262
17. Zhai WM, Wang KY, Lin JH (2004) Modelling and experiment of railway ballast vibrations. J Sound Vib 270(4–5):673–683
18. Punetha P, Nimbalkar S, Khabbaz H (2020) Analytical evaluation of ballasted track substructure response under repeated train loads. Int J Geomech 20(7):04020093
19. Punetha P, Nimbalkar S, Khabbaz H (2021) Simplified geotechnical rheological model for simulating viscoelasto-plastic response of ballasted railway substructure. Int J Numer Anal Meth Geomech 45(14):2019–2047
20. Punetha P, Nimbalkar S (2022) Performance improvement of ballasted railway tracks using three-dimensional cellular geoinclusions. Geotext Geomembr 50(6):1061–1082
21. Punetha P, Nimbalkar S (2023) An innovative rheological approach for predicting the behaviour of critical zones in a railway track. Acta Geotech 18:5457–5483
22. Punetha P, Nimbalkar S (2022) Geotechnical rheological modeling of ballasted railway tracks considering the effect of principal stress rotation. Can Geotech J 59(10):1793–1818
23. Farooq MA, Nimbalkar S, Fatahi B (2022) Sustainable applications of tyre-derived aggregates for railway transportation infrastructure. Sustainability 14(18):11715

Pavements of the Future: Decarbonisation, Digitalisation and Environmental Responsibility

Abir Al-Tabbaa

Abstract Many recent transportation-related strategy documents present roadmaps for how to deliver net zero through decarbonisation, digitalisation and environmental responsibility. This keynote lecture presents an overview of current strategies in the content of the UK highways and pavements highlighting the many challenges that we currently face and providing examples of current research initiatives that are tackling those challenges.

Keywords Pavements · Decarbonisation · Digitalisation · Environmental impact

1 Introduction

Many transportation-related strategy documents have emerged over the recent past to address the challenges associated with climate change, both in terms of mitigation and adaptation strategies with a focus on resilience and decarbonisation, and many include roadmaps for how to deliver net zero by 2050, see for example the UK National Highways documents [1–4]. Digitalisation has also been high on transportation agendas and seen as an enabler towards both zero harm targets and delivering net zero [5–7]. Sustainability and environmental protection have also become very high on the transportation agendas acknowledging the central role that our transport infrastructure can play in enhancing the ecosystem and contributing to a positive natural environment [8]. The UK's strategic road network is valued at more than 128 billion pounds, and it is expected that this amount will increase with the majority of the assets still being in service in 2050. The assets that make up the network are ageing, and it is now acknowledged that the maintenance and repair of the existing assets (as opposed to demolishing and rebuilding) are the most sustainable approach to maintaining its services.

A. Al-Tabbaa (✉)
Department of Engineering, University of Cambridge, Cambridge, UK
e-mail: aa22@cam.ac.uk

B. Indraratna and C. Rujikiatkamjorn (eds.), *Recent Advances and Innovative Developments in Transportation Geotechnics*,
https://doi.org/10.1007/978-981-97-8245-1_17

Within the context of road pavements, strategies refer to aspects of the design, construction, operation and maintenance and much of this relates to the conditions of the pavement and its materials. Materials are a significant contributor to pavement deterioration and carbon emissions [2] and are hence important contributors to both enhancing safety and reducing carbon emissions. Our current roads remain highly traditional, in their design, construction and material use. They hence continue to have a high carbon footprint, primarily due to the continued use of asphalt, cement/concrete and steel and lack of alternatives. There is also a lack of innovation and there is slow adoption of new designs and construction processes, e.g. off-site manufacturing and 3D printing. Roads are maintained in reactive traditional processes, where proactive maintenance and early preventative repairs are not usually endorsed, and this is linked to limited government expenditure on infrastructure repairs and upkeep in general, as well as limited use of digital platforms and solutions that can assist in prioritising and optimising processes. In addition, the construction and maintenance of roads usually lead to negative environmental impacts, including diversions and disruptions, significant consumption of natural resources and significant impact on the natural environment. As a result, current road schemes mainly address the minimisation of such negative environmental impact rather than actively targeting positive environmental benefits. Hence, our roads do not usually serve additional functionalities and dealing with significantly impactful challenges such as flooding, are usually separately designed for.

National strategy documents are beginning to consider many of these interconnected and interlinked aspects. Recent UK National Highways documents provided roadmaps for the delivery of net-zero pavement materials [9] with many more strategy and roadmaps recently produced for construction materials in general and cementitious and concrete materials in particular, which will have a significant impact on pavements [10, 11]. In addition, the need for digitalisation, digital technologies and data analytics is acknowledged and will enable significant advancements in material efficiency, cost savings and proactive pavement maintenance [12, 13]. The increased impact of climate change and resulting increased frequency of extreme temperatures, increased freeze–thaw cycles and increased flooding has been clearly witnessed and is having a significant impact on pavements around the world with accelerated deterioration and increased occurrences and severity of damage, e.g. potholes, cracking and spalling in pavements [14–17]. The expected deployment of driverless cars will impact the distribution of vehicle loadings on pavements resulting in changes in the type and extent of pavement deterioration expected and observed [18]. The important of protecting our natural environment, natural habitats and natural resources and providing nature-positive schemes is now considered part and parcel of any infrastructure scheme. All these considerations will have a significant impact on the design, construction, maintenance and repair of pavements including surface and subsurface layers with expected greater integration of both aspects for enhanced performance and resilience and to evolve to meet the challenges of climate change.

2 Pavements of the Future

It is clear that there are significant challenges in decarbonisation, digitalisation and environmental responsibility to be addressed in the design, construction, operation and maintenance of our roads and highways. Decarbonisation of our roads and pavement materials, with their high carbon emissions, will enable the significant reduction in the carbon footprint of highways projects. This will accelerate the efforts towards delivering net-zero road projects. Examples of recent research include the development and testing of low-carbon pavement materials (cement, concrete, aggregates, binders, additives, reinforcement); the development of self-healing, self-strengthening and self-immunisation; smart material use with material reduction; use of recycled materials for maximising the use of waste streams; carbon capture and storage within pavements and the wider road infrastructure; the development and use of detailed carbon calculators and whole-life performance modelling and analysis.

Digitalisation of our pavement materials including automation of processes and the increased use of sensors will enhance efficiencies, improve performance and generate a significant body of highly valuable data. This will improve our understanding of our pavement material and roads, enhance our performance and deterioration models and enable pavements to contribute to monitoring their own health and communicate with digital platforms, including digital twins, and be guided and optimised by data-driven approaches. Examples of recent research include the development of self-sensing materials, with conductive fillers and fibres as well as coatings; the deployment of carbon nanomaterials and graphene; the extensive use of different sensor and sensor devices, including innovative sensors for material health monitoring and early warning signals; the exploitation of digital tools; data capture, interpretation, interrogation and management as well as the development and testing of off-site, prefabrication and 3D printing manufacturing processes.

Environmental responsibility in roads and pavement materials will enable highway projects to play a central role in minimising negative environmental impacts of schemes and projects and enhancing the surrounding environment. Recent research has focused on designing pavement materials with added functionality, both technical and environmental, including reduction of water run-off, reducing air pollution and reducing noise; protecting the environment through the use of locally sourced materials, material reuse and delivering zero waste projects as well as future-proofing through the delivery of climate resilient pavements and the use of life cycle analyses.

3 Relevant Current Research Initiatives

The UK government has recently invested heavily in research and innovation initiatives and projects that focus on delivering advances in transportation, and in particular roads, and deliver the next generation of roads fit for the future. Examples of such major initiatives include Digital Roads Prosperity Partnership and the National Hub for Decarbonised Adaptable and Resilient Transport Infrastructure (DARe). Digital Roads is a £8.6M business-led collaborative research project between the University of Cambridge and Costain and National Highways, supported by the UK Engineering and Physical Sciences Research Council (EPSRC). Its aim is to lead a digital transformation of our vital road network to make them safer and greener. It explores how the four areas of smart materials, digital twins, data science and automation and robotics can be integrated to develop a connected physical and digital road infrastructure system. DARe is a £10.5M project, led by the University of Newcastle in collaboration with the Universities of Cambridge, Heriot-Watt and Glasgow and over 50 industry partners. It is funded jointly by EPSRC and the UK Department for Transport (DfT). It is a highly multi-disciplinary and aims to help upgrade and decarbonise the UK's complex and interconnected national, regional and local transport infrastructures. It will identify pathways and solutions for delivering a resilient net-zero transport system and will use and integrate models between assets and systems and use storylines and case studies to demonstrate effectiveness and value. Sharing knowledge, data and experience is at the heart of the DARe hub.

At a European level, the recently funded Future Roads COFUND Fellowship programme expands the remit of the Digital Roads Prosperity Partnership at the University of Cambridge. Future Roads is a HORIZON 2020 MSCA COFUND £5.9M project, co-funded by over 24 industry partners. It offers post-doctoral fellowships linked to the Digital Roads thematic areas of digital twins, smart materials, data science and automation and robotics and has a fifth theme of system-level sustainability. It addresses synergetic and complementary research areas and challenges to Digital Roads, expanding its boundaries and allowing fellows to propose their own research projects in responses to programme-identified challenges.

4 Examples of Current Research Challenges

Research challenges currently being addressed in the above research and aligned initiatives are presented here under the three main headings of decarbonisation, digitalisation and environmental responsibility where there is clear extensive synergies.

4.1 Recarbonisation

The UK Net-Zero Highways document [2] has set a target of net zero for maintenance and construction by 2040. Current road construction and maintenance projects are far from being carbon neutral with pavement materials being a major contributor to carbon emissions. Half of infrastructure carbon is associated with the maintenance and repair of assets throughout their service life. There is also a huge push to upgrade existing road assets in preference to building new ones. In addition, transportation is one of the most energy-consuming sectors yet significant opportunities lie with the road infrastructure for energy harvesting, including solar, mechanical (vibration and friction from vehicles) and geothermal energy. Hence, some of the challenges that are being posed include the following: Can suitable low-carbon or ultra-low-carbon and durable pavement materials be designed, validated and rolled out to contribute to carbon reductions in both construction and maintenance? Where within the different pavement materials and components can we make the quickest and largest impact and how? Can we rethink the design and performance of our pavement materials and structures to reduce material need and use and design more efficiently with less material consumption? Can we, and by how much, capitalise on the generation and implementation of recycled materials, waste-based materials and locally sourced materials? What and where are the most relevant and compatible national waste streams and where can we simultaneously help solve certain pressing waste and pollution problems? What are the opportunities for carbon capture and storage/sequestration within the pavement materials and the wider road infrastructure and network materials and assets. Where is the low-hanging fruit in the decarbonisation of road materials and is it in addressing capital carbon or operational carbon? Can we continue to reduce the carbon footprint of pavements throughout their service life? How do we quantify those benefits in whole-life carbon calculations and life cycle analyses? Can we develop sufficiently accurate carbon analysis tools? Can lessons be learnt from existing or past projects? What is the best way to balance cost, carbon and longevity? How can current carbon calculators be expanded and enabled to accurately and reliably assess the whole-life carbon of roads? How can we accelerate the validation of new low-carbon materials and products to assist with their implementation into specifications and standards? Could the high pavement temperatures witnessed this summer, leading to extensive pavement damage, be harvested as energy and at the same time reduce surface heat to deliver double benefits? What is the potential for material innovations both in energy capture and storage? What is the potential for, smart materials, e.g. piezoelectric, phase change or pyroelectric materials, thermochemical storage systems, highly reflective adhesive films, such as ethylene–vinyl acetate (EVA) films, conductive fillers, smart transparent coatings, layered composites and energy storage within concrete batteries in addressing this challenge? What are the cost and carbon values and environmental impact of such systems?

4.2 Digitalisation

The National Highways Digital Strategy [5] aims to deliver significant automation to the road network by 2025 and provides a roadmap towards this target. This commitment is paving the way for a range of innovative and impactful autonomous, robotic, remote and smart delivery approaches. Currently, automation in pavement practices is extremely limited. There is hence a huge potential for automation in the design, construction, inspection, maintenance and replacement of pavements. We also know that data play a fundamental role in enhancing our understanding of material and structural behaviour providing knowledge, insights and predictions that were not possible before. At the same time, the design requirements are changing to support the long-term vision of full digitalisation, connectivity and autonomy as well as delivery of net zero. Examples of the challenges currently being addressed here include the following: How can smart materials digitally enable proactive asset interventions and maintenance strategies? Can we design pavement materials that can look after their own state of health and access and provide live reporting and communication of this to enable timely and proactive asset management? How can pavement materials play a much more proactive role here? What are suitable and realistic mechanisms for digitising pavements and pavement materials? How can sensors, and other digital tools, and the integrated use of sensors in pavements and road assets, be used to predict when maintenance is needed? Could those sensors act as pavement materials heath monitoring indicators? Could the use of sensors enable proactive management of pavement drainage and run-off? Could we design or deploy low-cost multifunctional sensors, e.g. for combined asset management, traffic flow analysis and temperature measurements? Could pavement sensor technology be designed for the future network, e.g. compatible with electrification of the network? How can data that are currently available or which can be generated on-demand be used to deliver the data-driven smart pavements of the future? How can we facilitate the integration and seamless interaction between different data sources (evidence-based laboratory generated data, modelling data, sensor data and pavement and environmental data) be used to deliver data-driven solutions. How can be capture data, sufficient data, from pavement materials to enable us to fully understand, communicate and connect the road asset to the road system around it? Is it feasible to design and construct a fully automated pavement or deliver automated pavement maintenance? What is the role of precasting or 3D printing in delivering automated pavements? How can we design Digital Roads with materials and sensors that enhance the life span, longevity and durability of this critical infrastructure? How should design specifications be updated for resilience based on the available data and future predictive models? What will be the most relevant failure criteria for designing resilient roads in a changing climate? What are the relevant pavement failure mechanisms that will impact material durability with changing fleet pattern and increased demand?

4.3 Environmental Responsibility

While road schemes can have a significant negative impact on their surrounding environment, they also have a significant potential to positively impact on that environment and be delivered in an environmentally responsible fashion. The negative impacts arise from the consumption of natural resources and generation of wastes during the construction and maintenance of highways as well as disruptions to the natural environment. Examples of challenges that are being addressed include: Can

Future Roads minimise the consumption of natural resources, and can we use existing roads as a quarry for future projects? Can we up-cycle "waste" materials and what technologies would be required to enable this? How can we address the barriers to adoption of such materials and products? Can we develop accelerated ageing tests to aid with adoption? How can those new materials be better integrated into the standards? Can we use waste-based and locally sourced materials, and if so, how much? What and where are the most relevant national waste streams and where can we simultaneously help solve certain pressing waste problems? What are the cost, carbon and environmental impacts of such activities and is there a convincing and meaningful business case to make? Can smart materials assist in achieving efficient design, whereby we design with less materials and reduce overdesign in the maintenance of assets? Can we use available bodies of data to model the performance and deterioration of materials and, ultimately, their whole-life performance to inform intervention needs? What innovative processed and reworking can be deployed to convert excavated materials from earthworks, including 'fines' into value-added products tailored for highway projects? Can cost-effective, low-carbon and ideally waste-based stabilising additives and associated processing be developed and deployed here? Can waste from earthworks be used to generate a range of construction materials to enable different applications within road projects, such as supplementary cementitious materials, binders, aggregates, earthen construction materials and constructed elements, e.g. curbs, barriers, etc.? Can rapid testing, design and quality assurance procedures be developed to enable rapid commercial deployment of such products? Can we explore regional feedstock to minimise transport, costs and degradation of the material due to stockpiling? Can we plan sufficiently well in advance to use projects in the pipeline to optimise the exchange process? Can we use existing project data, currently within the supply chain, to inform optimisation processes and future decisions? How can we enable circular economy approaches within the network and beyond? What are the costs, carbon and environmental impacts/benefits of such activities and processes and is there a convincing and meaningful business case to make? Can we design quieter concrete roads and roads that can clean the air around them?

5 Selected Examples of Current Research Activities

This section presents an overview of current projects exploring some of the challenges above.

5.1 Self-healing of Cement-Stabilised Soils

Soil–cement systems are widely employed in transportation infrastructure, e.g. soil stabilisation, pavements and stabilised subgrade, embankments and slopes, cut-off walls and environmental remediation. For those systems, repeated freeze–thaw deterioration and damage are a major issue resulting in durability related problems. Recent research has explored the potential for augmenting those systems with self-healing capabilities. Self-healing of cementitious systems has emerged as a promising approach to extend the life of assets, although most focus so far has been on concrete and structural elements [19], rather than cemented-soil systems and geotechnical applications [20]. Given the much lower strength of those systems compared to concrete, the challenges associated with developing self-healing soil–cement systems are different. In addition, research work has also explored the potential for providing self-immunity to minimise or even eliminate the damage in the first place particularly under cyclic freeze–thaw conditions. Recent research has focused on exploring the potential for additives such as super-absorbent polymers (SAPs), microcapsules and magnesia pellets for self-healing and/or self-immunity in a range of soil–cement systems [21–25]. The microcapsules were specifically developed containing sodium silicate, as their cargo, that acts as a healing agent for cementitious systems [26]. Those microcapsules performed well in sealing and healing cracks, observed in recovered permeability values. Air entraining capsules as well as super-absorbent polymer pellets were found effective, through different mechanisms, in significantly enhancing resistance to freeze–thaw providing immunity leading to reduced, and in some cases, eliminated, damage.

5.2 *Self-healing Aggregates*

Aggregates are extensively used in all pavement layers, and the use of natural aggregates presents a significant draw on natural resources. Waste-based aggregates and the recycling of aggregates are two areas of research that have received considerable attention over the years with targeting increased dosages forming a current area of focus. Given the significant volume that aggregates occupy in soil–cement systems and as part of research on developing self-healing materials, it is advantageous to explore the use of the aggregate constituents, particular if they are porous, to house the healing agent. Work was carried out in which Lytag and expanded clay, both porous lightweight waste-based aggregates, were impregnated with sodium silicate and sealed with a polymer-based coating, to form a self-healing product [27–29]. The self-healing aggregates were used as partial replacement for sand. Permeability, water sorptivity, ultrasonic pulse velocity, strength and stiffness measurements were used to assess the self-healing performance in cracked soil–cement samples and showed significant promise. Successful large-scale testing demonstrated the scalability of the products and performance [30, 31], and numerical analyses were performed to optimise the aggregates dosage [32].

5.3 Self-healing and LCA of Roller-Compacted Pavement Concrete

Roller-compacted concrete (RCC) [33–36], which takes its name from the construction method used—placed by modified asphalt paving equipment, has the same constituents as conventional concrete but mixed in different ratios. With significantly less water used, the mix is much drier and has no slump and hence does not require formwork for placement and has no reinforcement. The strength of the finished material is reliant on the mix constituents and level of compaction achieved. RCC also has the potential to be trafficked significantly earlier than conventional Portland cement (PC) concrete. National Highways introduced RCC as a road base option in 2020. Construction with RCC is a relatively efficient and cost-effective process and has been shown to have a lower carbon footprint than continuous reinforced concrete pavements (CRCP) [37]. A cradle-to-gate life cycle analysis (LCA) was performed with a functional unit being one mile of a dual carriageway demonstrating that RCC pavements have less than half the environmental impact of CRCP due to lower thickness of RCC pavement lifts and lower quantity of PC compared to CRCP for the same design parameters [37]. The resistance of low-carbon RCCs in the presence of water and de-icing salts was explored [38, 39].

RCC with 80% ground granulated blast furnace slag (GGBS) content and SAPs and polypropylene (PP) fibres as additives was evaluated for compressive strength, freeze–thaw resistance and mass scaling resistance. The results indicate that both additives improved the compressive strength and freeze–thaw resistance of concrete with and without de-icing salts. The inclusion of PP fibres was found to be more effective than the addition of SAPs to mitigate the extent of internal structural damage and mass scaling of the concrete mixes with respect to the reference concrete after 56 freeze–thaw cycles.

5.4 Recycled Pavement and Excavation Materials

The recycling of pavements and excavation materials is currently a very hot topic with significant interest from industry and a general interest in facilitating the reuse of all components and fractions of recycled materials. Recent research has focused on waste fines (0–4 mm particle size) as there are currently no commercially viable techniques to facilitate their integration within products and there is significant lack of knowledge and understanding of their significant potential for added value. Such waste fines mainly arise from mining and quarrying, excavation activities in construction projects as well as in the washing and recycling of excavated materials, e.g. from road planning activities. Studies have explored the conversion of these fines, silts and clays, into fillers, cementitious materials for geopolymers and calcined clay cements and aggregates in pavements, as well as in producing artificial soils and in earthen construction [40–46]. Taking geopolymer and calcined clay cement applications as an example, a range of different calcination treatments have been investigated to enhance the reactivity as well as a range of different alkali activators. Although kaolin clays have the ideal mineralogy for calcined clay, field excavated clays usually vary depending on the location and include a range of different clay mineralogies. The feasibility of alternative clay mineralogies and blended compositions have hence been explored. A recent reactivity test, specifically aimed at clays, the R3 bound-water test, was developed by RILEM and is extensively used on those studies. The ability to use the significant volumes of these high value mineral wastes will provide significant environmental benefits,

5.5 Thermal Properties of Pavement Materials

Investigating the thermal performance of pavements is important in applications such as heating pavements (to prevent build-up of snow and icy road conditions), reducing the negative impact of freeze–thaw cycles, cooling pavements against urban heat island effect and exploring thermal energy capture and storage. Recent research activities have focused on heating roads, shallow ground source energy systems and the use of conductive fillers, fibres and phase change materials to optimise thermal performance [47–51]. A recent emerging technology is the use of heated ribbons that are embedded within the pavement for melting the snow and ice at the surface [47]. A mathematical thermal model using the 2D heat equation was used to explore the effect of variables and overall performance optimisation as well as local climates [47]. A limited number of field trials have been successfully performed exploring the upscaling and in situ performance of the technology. The technology is still at its infancy with current research focusing on the thermal profiling with varying configurations, geometry and spacing of the heated ribbons in order to optimise their design [48].

Other recent research has focused on enhancing the thermal properties of different soils with additives to enable temperature regulation in pavements to reduce the negative effect of extreme temperatures and temperature gradients as well as enhance the thermal capacity of the ground associated with shallow geothermal energy systems [49–51]. Conductive fillers, including copper filings and graphene, were explored in grouts for ground improvement application [49, 50] for optimising the thermal performance of thermo-active geostructures including pavements. Phase change materials are also being explored for their potential in thermal energy storage to help dissipate heat across pavements, measured using high-resolution thermal imaging cameras [51]. LCA analysis is also being carried out to demonstrate how ground source energy systems can be coupled with geostructures to deliver low-cost and low-carbon heating and cooling including in the context of transport infrastructure [52].

5.6 *Digital Image Correlation for Smart Pavement Damage Detection and Repairs*

Particle image velocimetry (PIV) or digital image correlation (DIC) developed for deformation measurements in soils for geotechnical applications [53] is currently being expanded and advanced for applications in pavement materials, using both stationary and mobile imaging systems, for detecting strains and damage initiation [54, 55]. DIC was explored to map the evolution of strains in concrete pavement samples subjected to compressive stresses [54]. Susceptibility to camera movements, shadows and changes in lighting are being explored by data scientists who are deploying techniques including automatic orthorectification of images and image stitching [54]. DIC was also coupled with perspective warping to measure material displacements or strains that can be an early warning of pavement failure [55]. Perspective warping enables the by-passing of the limitation of fixed DIC applications and allows for the collection of images captured at angles from mobile cameras such as everyday cars in motion. This provides new opportunities for the effective prediction of road damage formation and consequently predictive maintenance.

5.7 *Sensors and Data-Driven Approaches for Pavement Performance Monitoring*

Sensors and data-driven approaches are now being widely used in many applications, while they are still in their infancy in pavements related applications. Sensors can be deployed in pavements to provide real-time feedback on the performance and health of the road and provide data-based evidence to assist with informed decisions. Likewise, extensive bodies of data exist, in the form of traffic data, weather data, data from car sensors as well as road surveys, that are an extremely valuable source of information and become even more powerful when integrated providing insights that would not be arrived at otherwise. Such knowledge would enable development of proactive maintenance strategies and strategies for extending the life of our highway assets reducing cost, carbon and customer disruptions. Recent research has explored the applicability, robustness and value of a wide range of sensors that measure pavement properties such as strain, displacement, temperature, moisture, relative humidity, electrical resistance and electrochemical potential and how individually and collectively they can enhance our knowledge on material performance [56, 57]. Embedment of sensors in concrete beams, together with the use of a range of non-destructive testing, including ultrasonic pulse velocity and half-cell potential, has provided invaluable information on the robustness and longevity of those sensors as well as the reliability of the data [56]. Data are also an extremely valuable source when building pavement deterioration models and using those models to predict performance. Existing data-driven performance models developed globally for different types of pavements and various climates and environmental conditions have recently been reviewed and assessed for their capability to handle the time-dependent nature of the data and use of existing data to forecast future pavement conditions [58].

6 Conclusions

This keynote lecture explored challenges that face our future pavements in the context of decarbonisation, digitalisation and environmental responsibility. Examples of current research activities to address those challenges were presented under different headings including self-healing of cement-stabilised soils, self-healing aggregates, self-healing and life cycle analysis of roller compacted concrete, recycled pavements and excavation materials, thermal properties of pavement materials digital image correlation for smart pavements damage detection and repairs and sensors and data-driven approaches for pavement performance monitoring. These examples highlight the close interconnectivity between the different aspects of transportation geotechnics and pavement engineering within the context of pavement materials.

Acknowledgements The authors would like to acknowledge the many relevant research grants: EPSRC Digital Rods Prosperity Partnership (EP/V056441/1), EPSRC Resilient Materials for Life Programme Grant (EP/P02081X/1), Horizon 2020 MCSA Future Roads COFUND Fellowship programme (101034337), EPSRC SAFEGround: Sustainable, Flexible and Efficient Ground-Source Heating and Cooling Systems (EP/V042149/1), EPSRC Centre for Doctoral Training in Future Infrastructure and Built Environment: Resilience in a Changing World (EP/S02302X/1) and the EPSRC and Department for Transport funded National Research Hub for Decarbonised Adaptable and Resilience Transport Infrastructure (DARe) (EP/Y024257/1).

References

1. National Highways (2021) Connecting the country: our long-term strategic plan to 2050. https://nationalhighways.co.uk/media/ob1lqvqr/cre22_0150-masterplan-national-highways-ris3_final-1.pdf
2. National Highways (2021) Net zero highways: our 2030/2040/2050 plan. https://nationalhighways.co.uk/media/eispcjem/net-zero-highways-our-2030-2040-2050-plan.pdf
3. National Highways (2022) Preparing for climate change on the strategic road network—third adaptation report under the climate change act. https://nationalhighways.co.uk/media/z1ndodqx/preparing-for-climate-change-on-the-strategic-road-network.pdf
4. National Highways (2020) Delivery plan 2020–2025. https://nationalhighways.co.uk/media/vh0byhfl/5-year-delivery-plan-2020-2025-final.pdf
5. National Highways (2021) Digital delivery: enabling outcomes data and technology. https://nationalhighways.co.uk/media/prflg344/digital-delivery-enabling-outcomes-with-digital-data-and-technology-final-18-may-2021.pdf
6. National Highways (2023) Connected and autonomous plant—roadmap to 2035. https://assets.highwaysengland.co.uk/Innovation+Hub/CAP+roadmap.pdf
7. ASTM (2023) Advanced technologies for digitalisation of construction industry. https://amcoe.org/advanced-technologies-for-digitalization-of-construction-industry/

8. National Highways (2023) Environmental sustainability strategy. https://nationalhighways.co.uk/media/g5yfcl3m/nh-environmental-sustainability-strategy_final_020523.pdf
9. National Highways (2022) Net zero highways: our zero carbon roadmap for concrete, steel and asphalt. https://nationalhighways.co.uk/media/wmcie10p/net-zero-roadmap-for-concrete-steel-and-asphalt.pdf
10. Institution of Civil Engineers (2021) Low carbon concrete roadmap. https://www.ice.org.uk/media/q12jkljj/low-carbon-concrete-routemap.pdf
11. Mineral Products Association (2021) UK concrete and cement industry roadmap to beyond net zero. https://www.mineralproducts.org/MPA/media/root/Publications/2020/MPA-UKC-Roadmap-to-Beyond-Net-Zero_Oct20.pdf
12. National Highways (2021) Introduction to digital roads. https://nationalhighways.co.uk/media/2chotw13/introducing-digital-roads.pdf
13. National Highways (2021) Digital roads 2050 interactive roadmap. https://nationalhighways.co.uk/media/r3nn1h3m/digital-roads-2025-interactive-roadmap_final.pdf
14. The Independent Newspaper. The world's roads aren't ready for climate change. 22 Aug 2022. https://www.independent.co.uk/news/long_reads/cliamte-crisis-change-roads-melt-b2149899.html
15. Britain's pothole crisis roads are crumbling (March 2023). https://www.thesun.co.uk/motors/21781341/britains-pothole-crisis-roads-are-crumbling-repair-backlog/
16. Construction Magazine (2023) UK's pothole problem deepens (Jan 2023). https://constructiondigital.com/articles/uks-pothole-problem-deepens-will-we-be-able-to-keep-up
17. FHWA (2021) Impact of automated vehicles on highway infrastructure. https://www.fhwa.dot.gov/publications/research/operations/21015/21015.pdf
18. Merhebi GH, Joumblat R, Elkordi A (2023) Assessment of the effect of different loading combinations due to truck platooning and autonomous vehicles on the performance of asphalt pavement. Sustainability 15(14):10805
19. De Belie N, Gruyaert E, Al-Tabbaa A, Antonaci P, Baera C, Bajare D, Darquennes A, Davies R, Ferrara L, Jefferson T, Litina L, Miljevic B, Otlewska A, Ranogajec J, Roig-Flores M, Paine K, Lukowski P, Serna P, Tulliani JM, Vucetic S, Wang J, Jonkers HM (2018) A review of self-healing concrete for damage management of structures. Adv Mater Interfaces 5:1–28
20. Al-Tabbaa A, Harbottle MJ (2015) Self-healing materials and structures for geotechnical and geo-environmental applications. XVII European conference on soil mechanics and geotechnical engineering, Edinburgh
21. Chen JT (2019) Development and performance of self-healing and self-immune soil-cement systems subjected to freeze-thaw cycles. Ph.D. Thesis, University of Cambridge
22. Chen JT, Jin F, Al-Tabbaa A (2018) Preliminary investigation on the development and performance of self-immune and self-healing soil-cement systems under freeze-thaw cycles. Proceedings of the GeoShanghai international conference (GSIC), Shanghai, pp 84–91
23. Cao B, Chen J, Al-Tabbaa A (2021) Crack-resistant cement–bentonite cut-off wall materials incorporating superabsorbent polymers. Can Geotech J 58:800–810
24. Cao B, Souza L, Wang F, Xu J, Litina C, Al-Tabbaa A (2022) The first microcapsule-based self-healing cement-bentonite cut-off wall materials. Geotechnique 73(2):105–114
25. Cao B, Souza L, Xu J, Mao W, Wang F, Al-Tabbaa A (2021) Soil mix cut-off wall materials with microcapsule-based self-healing grout. J Geotech Geoenvironmental Eng 147(11):04021124
26. Kanellopoulos A, Giannaros P, Palmer D, Kerr A, Al-Tabbaa A (2017) Polymeric microcapsules with switchable mechanical properties for self-healing concrete: synthesis, characterisation and proof of concept. J Smart Mater Struct 26
27. Alghamri R, Kanellopoulos A, Al-Tabbaa A (2016) Impregnation and encapsulation of lightweight aggregates for self-healing concrete. Const Build Mat 124:910–921
28. Alghamri R, Kanellopoulos A, Litina C, Al-Tabbaa A (2018) Preparation and polymeric encapsulation of powder mineral pellets for self-healing cement based materials. Constr Build Mater 186:247–262
29. Alghamri R, Al-Tabbaa A (2020) Self-healing of cracks in mortars using novel PVA-coated pellets of different expansive agents. Constr Build Mater 254:110254

30. Alghamri R, Al-Tabbaa A (2022) Field investigation on the performance of smart aggregates for self-healing of concrete in different curing regimes. International conference—resilient materials for life, Sept, Cardiff, pp 206–210
31. Alghamri R, Rengaraju S, Al-Tabbaa A (2022) Large-scale laboratory trials of smart aggregates for self-healing in concrete under different curing regimes. Cem Concr Compos 104864
32. Alghamri R, Jahjouh M, Alastal K, Al-Tabbaa A (2022) A geometric study for determining the optimum volume fraction of pre-embedded capsules for self-healing of cracks in a cementitious matrix. Eng Res Express 4(3):035044
33. British Standards Institution (2019) Hydraulically bound materials for civil engineering purposes—specification for production and installation—BS9227. BSI
34. National Highways (2020) Design manual for roads & bridges, CD226 design for new pavement construction. National Highways
35. National Highways (2020) Manual of contract documents for highway works, volume 1—specification for highway works—series 1000. National Highways
36. Britpave (2022) Roller compacted concrete pavements-guide for design materials and construction. Britpave
37. Telford A, Rengaraju S, Mohammed H, Nassar A, Al-Tabbaa A (2024) Life cycle analysis of low carbon roller compacted concrete pavements. TRA conference Dublin, April 2024
38. Afroughsabet V, Al-Tabbaa A (2022) Influence of SAPs and GGBS on the freeze-thaw resistance of concrete pavement. Proceeding 8th international conference on self-healing materials, Milan
39. Afroughsabet V, Al-Tabbaa A (2022) Effect of SAPs and polypropylene fibres on the freeze-thaw resistance of low carbon roller compacted concrete pavements. Proc Int SMARTINCS'23 Conf, MATEC Web Conf 378:08006 (2023). https://doi.org/10.1051/matecconf/202337808006
40. Solouki A, Fathollahi A, Viscomi G, Tataranni P, Valdre G, Coupe S, Sangiorgi C (2021) Thermally treated waste silt as filler in geopolymer cement. Materials 14(17):5102. https://doi.org/10.3390/ma14175102
41. Solouki A (2022) Waste mineral filler recycling in new pavement solutions, Ph.D. Thesis. University of Bologna
42. Solouki A, Tataranni P, Tarsi G, Sangiorgi C (2023) Waste silt as filler in hot asphalt: a laboratory characterisation. Appl Sci 13(6):3473
43. Solouki A, Tataranni P, Sangiorgi C (2022) Thermally treated waste silt as geoploymer grouting material and filler in semiflexible pavements. Infrastructure 7(8):99
44. Solouki A, Viscomi G, Lamperti R, Tataranni P (2020) Quarry waste as precursors in geopolymers for civil engineering applications: a decade in review. Materials 13(14):3146
45. Harding PI (2023) Study of chemical and environmental performance of quarry waste kaolin residue fur use in earthen applications. MRes Thesis, University of Cambridge
46. Coppack T (2022) Clays: cements of the future. MEng Thesis, University of Cambridge
47. Adam Q, Segui P, Al-Tabbaa A, Bilodeau J-P (2024) Enhancing the performance of electrically-heated pavements with strips of thermally-conductive asphalt concrete: a numerical investigation. 14th International society for asphalt pavement, Montreal
48. Adam Q, Al-Tabbaa A (2024) Thermal performance optimization of ribbons-based electrically heated pavements: a numerical study. 5th international conference transportation geotechnics, Sydney
49. Cao B, Wang X, Al-Tabbaa A (2024) Improvement of soil thermal conductivity with graphite-based conductive cement grouts. ASCE J Geotech Geoenviron Eng (Submitted)
50. Cao B, Jin F, Rengaraju S, Al-Tabbaa A (2024) Climate resilience and energy harvesting of thermos-active roads. 5th international conference transportation geotechnics, Sydney
51. Alfalouji I (2023) Innovative thermal energy storage materials system in concrete pavements. First year report. University of Cambridge
52. Balwit-Cheung N, Beh M, Rengaraju S, Al-Tabbaa A (2024) Life cycle assessment of ground improvement methods to enhance the efficiency of thermo-active geostructures within transport infrastructure. 5th international conference transportation geotechnics, Sydney

53. Stanier SA, Blaber J, Take WA, White D (2016) Improved image-based deformation measurement for geotechnical applications. Canadian Geotech J 53(5):727–739
54. Palin D, Taha H, Vlachakis C, Rengaraju S, Hadjidemetriou GM, Marie d'Avigneau A, Anvo NR, Girolami M, Al-Tabbaa A (2024) Assessing road repair materials via digital image correlation. Transport research arena 2024, Dublin
55. Vlachakis C, Marie d'Avigneau A, Palin D, Hadjidemetriou GM, Taha HM, Rengaraju S, Anvo NR, Girolami M, Al-Tabbaa A (2024) Vision-based multi-perspective road pre-damage detection. Transport research arena TRA2024, Dublin
56. Rengaraju S, Vlachakis C, Afroughsabet A, Al-Tabbaa A (2022) Large-scale laboratory trials of self-healing technologies. Proceeding 7th international conference in self-healing materials, Milan
57. Al-Tabbaa A, Jefferson T, Paine K, Gardner D (2022) The future of infrastructure: resilient materials for life. https://api.repository.cam.ac.uk/server/api/core/bitstreams/d5925ea2-1568-43a6-893e-d2d2f9d21221/content
58. Wang ZZ, Al-Tabbaa A, Hakim B (2024) Data-driven pavement performance modelling: a short review. 5th international conference transportation geotechnics, Sidney

Geotechnical Engineering for Large Infrastructure Projects Such as Offshore Wind Farms

Britta Bienen

Abstract Offshore wind farms are not only major infrastructure projects in themselves, but also tend to require significant investment in related infrastructure such as port facilities. Geotechnical engineering underpins the safe, reliable design and operation of all these infrastructures. Australia must build 41 GW of offshore wind power by 2040 to reach its CO_2 emission targets. To put this into context, the installed capacity in 2022 was 64 GW—globally, and there is not a single offshore wind turbine installed in Australia to date. Foundations have grown with the size of offshore wind turbines and are large structures that have to be accommodated at the marshalling port, be ensured to be installed to target depth offshore—often in complex seabed conditions—before withstanding millions of load cycles over typically 30 years of wind turbine operation offshore. This paper focuses on new solutions of rapid pile installation with low acoustic emissions and provides an outlook with a view of Australia's challenging seabed conditions.

Keywords Offshore wind · Piles · Installation · Cyclic loading

B. Bienen (✉)
Centre for Offshore Foundation Systems, University of Western Australia, Perth, WA 6009, Australia
e-mail: britta.bienen@uwa.edu.au

B. Indraratna and C. Rujikiatkamjorn (eds.), *Recent Advances and Innovative Developments in Transportation Geotechnics*,
https://doi.org/10.1007/978-981-97-8245-1_18

1 Introduction

1.1 Context

Offshore wind farms are large-scale energy infrastructure developments. Examples include Hornsea and Dogger Bank in the North Sea, which are constructed in stages and feature approximately 300 turbines each when complete. One rotation of a turbine can power a household for two days. These two offshore wind farms have an installed capacity of 6.2 GW, enough to power 8.4 million homes. The Dogger Bank offshore wind farm will be the UK's single largest source of energy [1].

Australia must build 41 GW of offshore wind power by 2040 to reach its CO_2 emission targets. To put this into context, the installed capacity in 2022 was 64 GW [2]—globally, and there is not a single offshore wind turbine installed in Australia to date.

In 2022, 8.8 GW of new offshore wind was added globally, with the Asia–Pacific region reaching 34 GW and becoming the world's largest offshore wind market [2]. With a double-digit growth rate of 15% and strong growth predicted in large emerging markets, 130 GWs of new offshore wind capacity are expected to be added in the next five years globally [2]. Floating offshore wind is reaching commercial scale, which opens immense potential in locations where offshore wind development would have been unimaginable until very recently.

China targets 100 GW by 2025, 200 GW by 2030, and 1000 GW by 2050. South Korea's 10th Basic Plan, released in December 2022, targets 14.3 GW by 2030 [2]. In Europe, energy ministers from the nine members of the North Seas Energy Cooperation (NSEC) agreed to reach at least 260 GW of offshore wind capacity by 2050. Figure 1 illustrates the required growth in global offshore wind capacity to reach net zero by 2050.

Offshore wind also opens up the opportunity to scale the production of green hydrogen and power-to-X, thereby aiding in the decarbonisation of energy intensive sectors. As a major source of green jobs, from manufacturing of turbine components through to installation and operation, offshore wind is increasingly seen as a positive

Fig. 1 Offshore wind capacity required to reach net zero by 2050 [3]

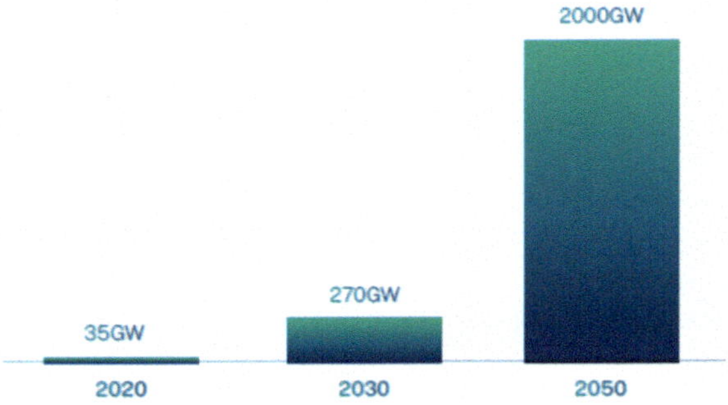

economic driver GWEC [2] providing tens of thousands of jobs (e.g. [4]), including in regional areas, which is also recognized by the Australian government [5] and should be an incentive to move the discussion away from a narrow focus on the lowest possible price. The Star of the South offshore wind farm project is set to invest about $10bn into Victoria and could provide about 8000 jobs over its 30-year lifetime.

In Australia, six priority areas for offshore wind have been identified [6], and applications for feasibility licences have closed for two of these areas, with announcement of the outcomes for the first offshore wind leasing round in the Gippsland area said to be imminent. Feasibility licences are valid for seven years in which time necessary work is carried out, including to obtain environmental and other approvals. A commercial licence is required to be obtained before construction can begin. Construction of offshore wind farms requires appropriate port and other related infrastructure, which may require significant and timely investment to enable project delivery. Geotechnical engineering underpins the safe, reliable design and operation of all this infrastructure.

1.2 Foundations and Ports

The latest turbines have a rated output of 15 MW, with rotor diameters exceeding 200 m. OWTs are complex systems with the majority of uncertainties attributed to the foundation–soil interaction (e.g. [7]). Foundations account for ~ 25% of the cost of offshore wind developments which are in the order of $10 billion per project.

The vast majority of offshore wind turbines have been installed on monopiles, which will also be the focus of this paper, noting that other low acoustic emissions solutions include suction bucket jackets and a range of anchoring options are being considered for floating wind developments. Foundations have grown with the size of offshore wind turbines, with the length of each rotor blade of the Haliade-X turbines used at the Dogger Bank offshore wind farm, for instance, being 107 m [8] and the turbine height of 260 m, almost twice the height of the London Eye [9] and similar in height to the Crown Sydney tower. These large parts of the wind turbine structure have to be accommodated at the port from which the wind farm construction is undertaken.

The port of Nigg (Fig. 2), for example, offers extensive laydown and storage areas adjacent to the quay, as well as project offices and other services.

To mention another example from overseas, the Port of Esbjerg Port (Fig. 3) on the southwest coast of Denmark was established as a fishing industry port in 1868. Today, it has been transformed through an infrastructure expansion process into the world's largest wind energy base port. Approximately 80% of the offshore wind capacity installed in Europe was shipped out from the Port of Esbjerg [10], despite Esbjerg having a population of approximately 110,000 (less than half of Wollongong or Newcastle). Since 2001, 23.6 GW of offshore wind turbines have been shipped from the port. Offshore wind is one of Denmark's most significant industries with

Fig. 2 Port of Nigg, UK (*Source* https://nigg.com/, 2023)

an annual turnover of more than $23 billion, and a large part of the Danish offshore wind supply chain are located in the Esbjerg hinterland [11]. In Australia, ports will need to be upgraded to enable offshore wind farm construction.

Fig. 3 Port of Esbjerg, Denmark

1.3 This Paper

This paper will draw from developments in foundations for offshore wind turbines, including new solutions of rapid pile installation with low acoustic emissions, and provides an outlook with a view of Australia's challenging seabed conditions.

2 Foundation Installation with a Focus on Low Acoustic Emissions' Solutions: Vibro-driven Piles

Impact-driven pile installation has a percussive acoustic signature with high magnitude peaks, requiring mitigation in some jurisdictions which typically takes the form of (double) bubble curtains through the water column but can also be mitigation of acoustic emissions at the source. Vibro-driving (Fig. 4) is an alternative low acoustic emissions pile installation method which does not necessitate noise mitigation measures. Reference [12] provide an overview of the acoustic signature for impact- and vibro-driven piles and radiation of the acoustic waves through the water column and seabed.

Irrespective of the pile installation method though, driveability needs to be addressed, the driving equipment selected and the pile sized well before installation can take place offshore. This is to ensure that the pile can be installed to target depth, which underpins pile performance post installation. It is therefore critically important to be able to accurately predict the soil resistance to pile driving (or vibration), including prediction of refusal (where the pile does not significantly advance) but also to anticipate risk of pile run (rapid uncontrolled penetration). Implications of such rapid penetration are mitigated in vibro-driven installations as the pile is held by clamps, and the vibro-driver is connected to the crane via cables. However,

Fig. 4 Vibro-driven pile installation offshore (*Source* Cape Holland)

several incidents have been reported of pile runs of impact-driven piles. The installation method also has implications for the remaining fatigue life of the pile post installation.

Vibro-driveability predictions have been proposed, including models based on force equilibrium (e.g. [13]). Combination with analysis of the vibro-driven pile motion allows models to also provide an estimate of the driving velocity (e.g. [14]) with few opportunities to date for validation against large diameter cylindrical monopiles (e.g. [15, 16]) as used for offshore wind farms. Evaluation of the vibrator-pile motions using a wave equation analysis further enables accurate estimate of the maximum stress level in the pile [17].

An opportunity arose to obtain a blind prediction using the Hypervib model (which was originally developed by Holeyman [18]) against physical data from carefully controlled centrifuge model tests of a thin-walled cylindrical pile vibro-driven into saturated sand [19]. The information provided included the cone tip resistance profile as obtained from CPTs performed in the centrifuge sample, pile dimensions as well as static forces due to self-weight of the pile and vibro-driver and the dynamic force exerted by the vibro-driver at the given frequency. Importantly, the experimental results were not shared with the prediction team. The results are compared in Fig. 5, with the red curve showing the physical measurements, while BE and HE denote the best estimate and high estimate predictions, respectively. The comparison is encouraging as the actual driving time remained between the best and high estimates, and importantly, the high estimate matches reasonably well the refusal depth obtained by the centrifuge test at 3.9 m.

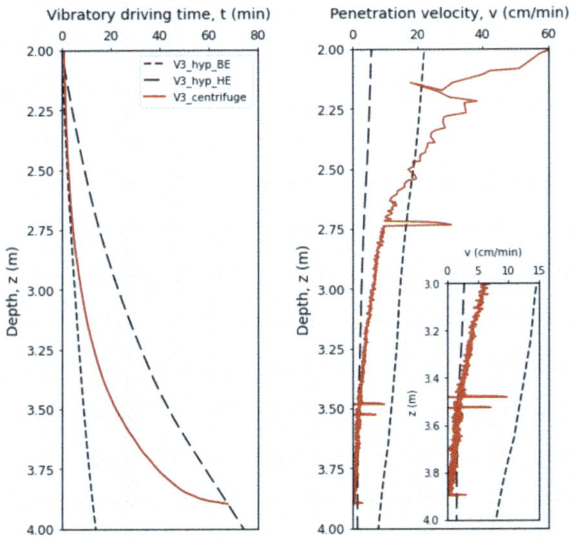

Fig. 5 Vibro-driven pile installation: measurements from centrifuge testing compared with prediction [19]

Further data relevant to vibro-driven installation of large diameter offshore piles would clearly be beneficial in improving confidence in predictions and fine-tuning both the methods and parameter values.

Large-scale onshore tests can utilize the same installation apparatus as offshore installations, with confidence in the mechanisms that govern the measured response being applicable to the field more generally. Recent examples of vibro-driven piles have been reported, e.g. in [20] and [21], which have provided insights into the lateral response of vibro-driven monopiles, including comparison with impact-driven piles installed at the same site. However, soil at onshore test sites is not usually fully saturated over the entire pile embedment length, which may influence (vibro-driven) pile installation and can lead to gapping under lateral loading. Gapping is not expected in the fully saturated conditions offshore. Gapping during cyclic lateral loading features in the field test results presented in [22], which includes detailed discussion and transfer of the proposed modelling to a fictitious fully saturated site without gapping.

Smaller scale testing allows experiments to be performed under controlled conditions in the laboratory. Reference [23] showed in tests with pile diameters up to 0.61 m that the stress development during pile installation in the vicinity of the pile wall can vary significantly not only between impact-driven and vibro-driven pile installations, but also depending on the way the vibro-driven pile installation was performed. This is because different combinations of static and dynamic forces result in different cyclic movements of the pile. The concept of the magnitude of reversal being influential in vibro-driving is not new [14, 24–27]. The complexity arises from the cyclic pile displacement being the outcome of the load-controlled process involving static and dynamic forces as well as frequency, with friction fatigue (e.g. [28]), dynamic effects [29–31] and the resulting drainage regime in saturated soil also playing a role.

Development of a mini vibro-driver [32] has enabled vibro-driven pile tests to be performed in a geotechnical centrifuge, which takes advantage of small-scale testing under controlled conditions with the experiments importantly performed at stress levels relevant to the field. The data (incl. results shown in Fig. 5) further contribute to the effort of obtaining physical evidence of vibro-driven pile installation, with systematic variation in the vibro-driving input aimed at advancing scientific understanding of the underlying mechanisms with the ultimate goal to provide recommendations for vibro-driving. Techniques such as digital image correlation of photos taken during vibro-driven installation of a half model pile against a transparent window (e.g. [33]) can provide further insights including sand density changes around the pile as well as physical data for validation of numerical models.

Numerical modelling can reveal changes in soil state, including changes in the stress field. This technique has been used to illustrate changes in soil state due to impact-driven pile installation (e.g. [34–36]) and is now applied to vibro-driven pile installations (e.g. [37]), with Fig. 6 providing an example of resulting changes from a cylindrical pile being driven into saturated sand, comparing effects between impact and vibro-driven installations.

Reference [38] also used numerical modelling in Class A predictions of field data from the North Sea obtained in the VISSKA project, which showed encouraging

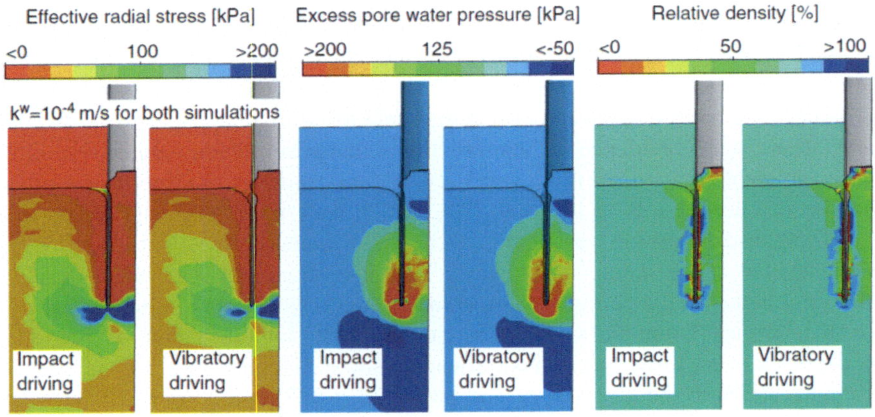

Fig. 6 Numerical modelling revealing changes to the soil state resulting from a 4 m diameter pile driven to 10 m depth in dense sand with a hydraulic conductivity of $k = 10^{-4}$ m/s [37]

results despite simplifying assumptions of homogenous soil conditions. Numerical modelling of the installation of thin-walled piles into saturated sand (which may only be four pile diameters but 400 times the wall thickness) is still a very complex boundary value problem.

3 In-Service Performance: Initial Stiffness and Response Under Large Numbers of Load Cycles

Research on the effect of the installation process on the subsequent lateral response of impact-driven piles has highlighted the importance of understanding the post installation soil state which governs the lateral response (e.g. [34–36]), with significant differences in both initial stiffness and ultimate capacity possible. For bottom-fixed offshore wind turbines, 0.5° degrees of rotation is often cited as the serviceability limit criterion. This includes installation tolerances and in-service performance over the lifetime which is typically 30 years.

Figure 7 compares the lateral response of piles loaded laterally following different installation processes. These results were obtained through numerical modelling. Figure 8 compares pile response to lateral loading measured in field tests in Altenwalde near Cuxhaven, Germany, from pairs of impact and vibro-driven installations [20].

Two of three vibrated piles were installed with vibration parameters selected to achieve high penetration velocity, i.e. to minimize the duration of pile installation. These piles behaved softer than the impact-driven piles under horizontal loading. One of the vibro-driven piles unintentionally had a much longer installation time and behaved differently. This pile showed only slightly softer behaviour compared with an

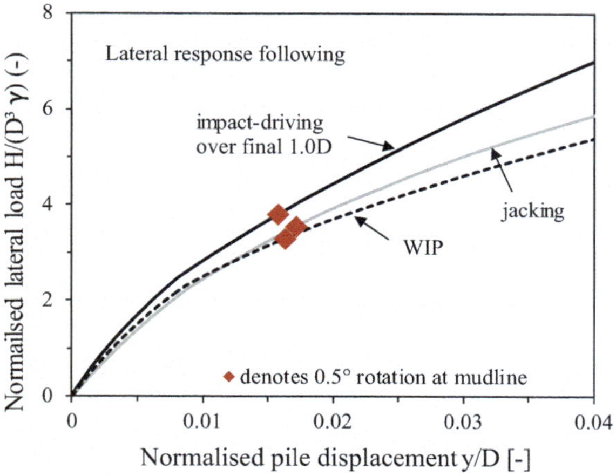

Fig. 7 Effect of pile installation (impact-driving, jacking, wished-in-place WIP) on the lateral response of an 8 m diameter monopile in dense Karlsruhe fine sand [34]

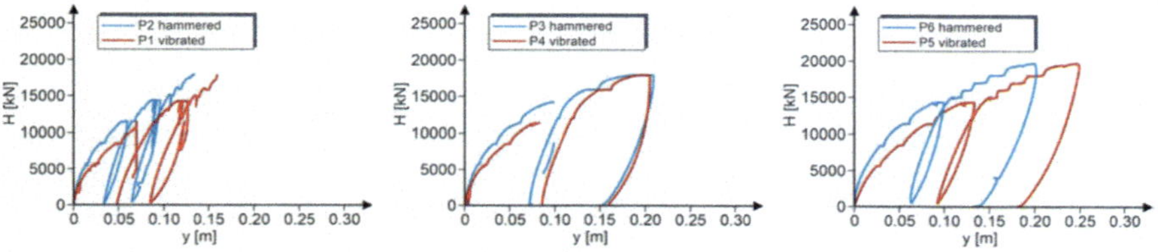

Fig. 8 Load–displacement responses measured in field tests [20]

impact-driven pile, strengthening the expectation that the vibro-driving parameters (static and dynamic forces as well as frequency) influence the resulting soil state which underpins pile performance post installation.

Reference [37] compare the lateral response of piles subjected to large numbers of load cycles (Fig. 9), with the results suggesting that the pile installation processes modelled in these numerical simulations yielded post installation soil states that resulted in negligible difference in the pile response following impact-driving (ID) or vibro-driving (VD), which aligns with the respective post installation soil states shown in Fig. 6. However, neglecting the effects of the pile installation process by commencing lateral cyclic loading from wished in place (WIP) conditions resulted in a softer response with larger pile head rotation.

These results were obtained employing an approach whereby modelling cycle-by-cycle for a small number of load cycles (here the first two cycles) is combined with extrapolation of the trend of the response over large numbers of cycles (High Cyclic Accumulation or HCA model, [39]). The HCA model has been validated against an extensive database of soil element test results (e.g. [40]) and boundary

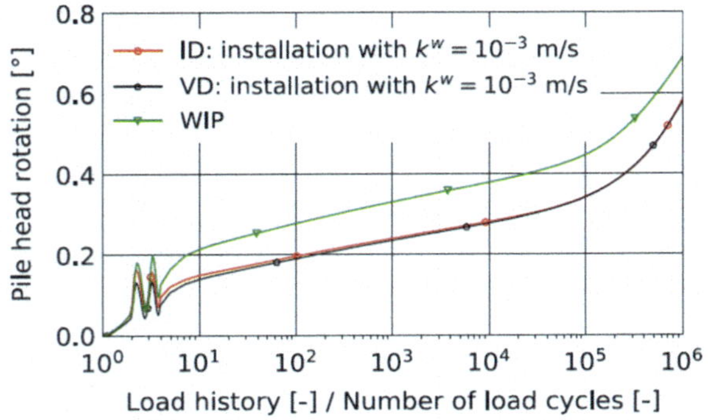

Fig. 9 Accumulated pile head rotation over a large number of lateral load cycles for vibro-driven (VD), impact-driven (ID) and wished-in-place (WIP) piles in medium dense sand [37]

value problems (e.g. [36, 37]). Parameters will need to be calibrated for seabed soils in regions with significantly different soil behaviour.

4 Conclusions and Outlook

Infrastructure is central to functioning societies around the world, and with global efforts to decarbonize required to gather pace, offshore wind farms and the ports that service their construction, operation, and maintenance play a key role. This paper has shone a spotlight on the geotechnical engineering underpinning these large infrastructure projects. Vibro-driving is a promising low acoustic emissions technique for pile installation, with limited experience to date related to larger diameter cylindrical piles like monopiles used offshore. The paper has discussed recent advances in scientific understanding of the mechanisms underpinning vibro-driving, which flow into development of prediction methods. Pile design and fabrication, as well as vessel and vibro-driver selection take place well ahead of installation offshore—all relying on accurate prediction of vibro-driveability. The installation process itself changes the soil state, and different vibro-driving conditions can lead to different outcomes which manifest themselves in the lateral response of the pile post installation. This affects both the initial stiffness and the performance under cyclic loading. Further research is clearly required and ongoing to enable recommendations to be formulated. Ideally, this can provide guidance on how to vibro-drive to influence the pile installation time and/or the post installation lateral response as required.

The expected seabed soils offshore Australia predominantly comprise carbonate sandy sediments which tend to exhibit substantial volume change when compressed and sheared, which is exacerbated by the crushable grains. These are the types of sediments that resulted in $340 million of remediation cost at the North Rankin A platform on Australia's North West Shelf where piles largely free-fell to just over 100 m depth—an expensive lesson learnt for pile design in carbonate seabeds. This was due to significant over-estimation of the foundation capacity when directly using design methods that were developed in siliceous sands that proved to be unconservative for the Australian carbonate seabed. Significant expertise and experience now exist with a focus on ultimate limit state (ULS) considerations related to oil and gas infrastructure, which is very different from the low strain response typical for monopiles supporting OWT over large numbers of load cycles. Little is known about the response of carbonate sands interacting with piles that are subjected to very large numbers of lateral load cycles—an area of exciting ongoing research.

This paper has focused on vibro-driven piles. Other low acoustic emissions foundation and anchoring solutions include suction buckets and helical piles.

Acknowledgements This work forms part of the activities of the Centre for Offshore Foundation Systems (COFS), in the Oceans Graduate School, and is the result of collaboration with colleagues at COFS and NGCF technical staff at UWA as well as colleagues in academia and industry around the world. This research is supported through the Australian Research Council (ARC) Discovery Project 200103466 "Understanding vibratory pile installation in sand and its effects on in-service pile response under lateral loading". This work was supported by resources provided by the Pawsey Supercomputing Research Centre with funding from the Australian Government and the Government of Western Australia.

References

1. Dogger Bank (2023). https://doggerbank.com/
2. Global Wind Energy Council (GWEC) (2023) Global wind report 2023
3. Global Wind Energy Council (GWEC) (2021) Global wind report 2021
4. Offshore Wind Industry Council (2022) UK offshore wind skills intelligence report. https://www.renewableuk.com/news/608235/New-report-shows-jobs-in-UK-offshore-wind-industry-to-grow-to-100000.htm
5. Department of Climate Change, Environment, Energy and Water (DCCEEW) (2023a). https://www.dcceew.gov.au/sites/default/files/documents/oct-budget-2022-23-jobs-fs.pdf
6. Department of Climate Change, Environment, Energy and Water (DCCEEW) (2023b). https://www.dcceew.gov.au/energy/renewable/offshore-wind
7. Kallehave D, Byrne BW, LeBlanc Thilsted C, Mikkelsen KK (2015) Optimization of monopiles for offshore wind turbines. Philos Trans Royal Soc. https://doi.org/10.1098/rsta.2014.0100
8. GE (2023) https://www.ge.com/renewableenergy/wind-energy/offshore-wind/haliade-x-offshore-turbine
9. Equinor (2023). https://www.equinor.com/magazine/how-we're-building-dogger-bank-wind-farm
10. Port of Esbjerg (2023) https://portesbjerg.dk/en/business-areas/offshore-wind
11. Oceanex Energy (2022) Energising Australia with offshore wind: supply chain and investment opportunities to be created by ocenex energy kickstarting a floating offshore wind industry in new South Wales. https://oceanexenergy.com/about-oceanex-energy/
12. Tsouvalas A, Peng Y, Molenkamp T, Sertlek O (2022) Environmental vibrations and noise due to offshore installation of foundation piles. Proceeding 11th international stress wave conference, Rotterdam, The Netherlands
13. Jonker G (1987) Vibratory pile driving hammers for pile installations and soil improvement projects. Offshore technology conference, Houston, USA, OTC5422
14. Dierssen G (1994) *Ein bodenmechanisches Modell zur Beschreibung des Vibrations-rammens in körnigen Böden*. Doctoral Thesis, University of Karlsruhe, Germany

15. Holeyman A, Whenham V, Peralta P, Ballard J-C, Pulukul SC (2020) Vibratory installation study of a monopile in dense sand. Proceedings 39th international conference on ocean, offshore and artic engineering, OMAE2020-19204, Florida, USA
16. Van Dorp R, Moscoso N, Bielefeld M, Verbeek G (2019) Prediction and monitoring of installation of offshore foundation monopiles for windfarms. Offshore technology conference, OTC-29400-MS
17. Holeyman A, Whenham V (2017) Critical review of the hypervib1 model to assess pile vibro-drivability. Geotech Geol Eng 35:1933–1951
18. Holeyman A (1993) HYPERVIB1 - An analystical model-based computer program to evaluate the penetration speed of vibratory driven sheet piles. Research report for BBR1, 23 p
19. El Haffar I, Ballard JC, Holeyman A, Hein Mazutti J, Bienen B, Randolph MF, Bransby MF (2023) Quantifying the vibro-drivability of monopiles in sand: prediction v. experimental results. Proceedings 9th international offshore site investigation and geotechnics conference: innovative geotechnologies for energy transition (OSIG), London, UK
20. Achmus M, Schmoor KA, Herwig V, Matlock B (2020) Lateral bearing behaviour of vibro- and impact-driven large-diameter piles in dense sand. Geotechnik 43(3):147–159
21. Metrikine AV, Tsouvalas A, Segeren MLA, Elkadi ASK, Tehrani FS, Gómez SS, Atkinson R, Pisanò F, Kementzetzidis E, Tsetas A, Molenkamp T, van Beek K, De Vries P (2020) GDP: a new technology for gentle driving of (mono) piles. Proceeding 4th international symposium on frontiers in offshore geotechnics (ISFOG), Austin, Texas, USA
22. Kementzetzidis E, Pisanò F, Tsetas A, Metrikine AV (2023) Gentle driving of piles at a sandy site combining axial and torsional vibrations: quantifying the influence of pile installation method on lateral behavior. J Geotech Geoenviron Eng (ASCE) 149(11). https://doi.org/10.1061/JGGEFK.GTENG-11179
23. Fischer J, Stein P (2022) Impact and vibratory driven pipe piles: the difference in soil stresses and bearing capacities due to the installation method. Proceedings 11th international stress wave conference, Rotterdam, The Netehrlands
24. Cudmani R (2001) *Statische, alternierende und dynamische Penetration in nichtbindigen Böden*. Dissertation, Heft 152, Veröffentlichungen des Instituts für Bodenmechanik und Felsmechanik, Karlsruher Institut für Technologie
25. Labenski J (2020) *Untersuchungen zum Einbringverhalten und dem lateralen Tragverhalten unter monotoner Einwirkung von in nichtbindigen Böden vibrierend installierten Monopiles*. Dissertation, Institut für Geotechnik der Universität Stuttgart
26. Massarsch KR, Weräll C, Fellenius BH (2020) Vibratory driving of piles and sheet piles—state of the practice. Proceeding institution of civil engineers—geotechnical engineering, paper 2000127
27. Rodger AA, Littlejohn GS (1980) A study of vibratory driving in granular soils. Géotechnique 30(3):269–293
28. Moriyasu S, Kobayashi S, Matsumoto T (2018) Experimental study on friction fatigue of vibratory driven piles by in situ model tests. Soils Found 58(2018):853–865
29. Massarsch KR, Fellenius BH, Bodare A (2017) Fundamentals of the vibratory driving of piles and sheet piles. Geotechnik 40(2):126–141
30. Viking K (2002) Vibro-driveability—a field study of vibratory driven sheet piles in non-cohesive soils. Ph.D. Dissertation, Department of Civil and Architectural Engineering Royal Institute of Technology (KTH), Stockholm, Sweden
31. Whenham V, Holeyman A (2012) Load transfers during vibratory driving. Geotech Geol Eng 30(5). https://doi.org/10.1007/s10706-012-9527-0
32. Hein Mazutti J, Bienen B, Bransby MF, Randolph MF, Wager G (2024) Development of a mini vibro-driver for pile testing in the centrifuge. Int J Phys Model Geotech. https://doi.org/10.1680/jphmg.23.00056
33. Remspecher F, Le VH, Rackwitz F, Herwig V, Matlock B (2019) Vibratory driven installation of monopoles—an experimental investigation of the soil-pile interaction. Proceedings 1st Vietnam symposium on advances in offshore engineering, Hanoi, Vietnam

34. Bienen B, Fan S, Schröder M, Randolph MF (2021) Effect of the installation process on monopile lateral response. Geotech Eng, Themed Issue Numer Model Large Strain Deformation 174(5):530–548. https://doi.org/10.1680/jgeen.20.00219
35. Fan S, Bienen B, Randolph MF (2021a) Effects of monopile installation on subsequent lateral response in sand, Part I: pile installation. J Geotech Geoenvironmental Eng (ASCE). https://doi.org/10.1061/(ASCE)GT.1943-5606.0002467
36. Staubach P, Machaček J, Moscoso MC, Wichtmann T (2020) Impact of the installation on the long-term cyclic behaviour of piles in sand: a numerical study. Soil Dyn Earthq Eng. https://doi.org/10.1016/j.soildyn.2020.106223
37. Staubach P, Machaček J, Bienen B, Wichtmann T (2022) Long-term response of piles to cyclic lateral loading following vibratory and impact driving in water-saturated sand. J Geotech Geoenv Eng. https://doi.org/10.1061/(ASCE)GT.1943-5606.0002906
38. Giridharan S, Stolle D, Moormann C (2023) Numerical prediction of the installation of vibratory monopile foundations for offshore wind energy projects. Proceedings 10th European conference on numerical methods in geotechnical engineering (NUMGE), London, UK
39. Niemunis A, Wichtmann T, Triantafyllidis T (2005) A high-cycle accumulation model for sand. Comp Geotech 32(4):245–263
40. Wichtmann T, Niemunis A, Triantafyllidis T (2015) Improved simplified calibration procedure for a high-cycle accumulation model. Soil Dyn Earthq Eng 70(3):118–132
41. Fan S, Bienen B, Randolph MF (2021b) Effects of monopile installation on subsequent lateral response in sand, Part II: lateral loading. J Geotech Geoenviron Eng (ASCE). https://doi.org/10.1061/(ASCE)GT.1943-5606.0002504

Experimental Study on the Biodegradation in Xanthan Gum-Treated Sandy Loam and Mitigation Using Biochar

Xueyu Geng

Abstract Biopolymers are eco-friendly materials used to enhance soil mechanical strength. However, due to the degradation effects of microbials, the strength of biopolymer-reinforced soil tends to attenuate over time. To address this issue, this study introduces another bio-based green material, biochar, to reinforce the soil in combination with the biopolymer. Static and dynamic triaxial tests were conducted to investigate the reinforcement effects. Water-soluble carbohydrates and β-glucosidase activity, along with soil's internal friction angle and cohesion, were measured after curing samples for 1, 7, 14, 21, and 28 days to quantify biodegradation. A scanning electron microscopy (SEM) test was performed to illustrate microstructure changes due to biodegradation and to reveal how biochar mitigates the degradation process in biopolymer-treated soil. It was found that xanthan gum forms hydrogels through hydrolysis reactions and establishes hydrogel bonds among soil particles, enhancing the soil's static and dynamic strength. However, microbes can degrade these hydrogels, resulting in a significant attenuation of soil reinforcement. Due to biochar's porous morphology and hydrophilic functional groups, the free water in the soil is gradually released and participates in the hydrolysis of xanthan gum. Additionally, hydrogels are captured by the voids in the biochar, leading to stronger inter-particle binding. Consequently, the strength of XG-treated soil exhibits a trend of slow growth over time. Biochar's angular structure enhances interlocking and embedding among particles, leading to a 42% increase in shear strength compared to untreated soil, and a 21% increase compared to soil treated with xanthan gum alone. Therefore, the combined approach of xanthan gum and biochar is advisable for reinforcing soil in engineering practice.

Keywords Soil reinforcement · Biopolymer · Biochar · Biodegradation

X. Geng (✉)
University of Warwick, Coventry CV4 7AL, UK
e-mail: Xueyu.Geng@warwick.ac.uk

B. Indraratna and C. Rujikiatkamjorn (eds.), *Recent Advances and Innovative Developments in Transportation Geotechnics*,
https://doi.org/10.1007/978-981-97-8245-1_19

1 Introduction

The quality of railway earthworks is critical to ensuring a safe and comfortable journey and the overall efficiency of the network. Undesirable ground conditions, such as low bearing capacities, low shear forces, and low permeabilities [1], are unavoidable when constructing these railway networks. The combination of climate change and dynamic cyclic loadings caused by moving vehicles makes construction and maintenance even more challenging [2]. The current best practice, such as grouting, heavily depends on cement, leading to a significant release of embedded CO_2, thereby causing a harmful environmental impact and exacerbating climate change even more.

Bio-inspired ground improvement methods have been studied extensively over the past decades to address environmental concerns. These methods are viewed as more environmentally friendly and sustainable alternatives to conventional practices [3]. Recently, biopolymer-treated soil has been known to have a comparable engineering performance because of the improvement in soil's static strength [4, 5]. The viscous biopolymer hydrogels formed by the reactions among biopolymer, water, and soil particles have been reported to exhibit noticeable impacts on reinforcing soil's unconfined compressive strength [6], shear resistance [7], and vegetation growth [8]. However, as a biodegradable material, biopolymers can undergo biodeterioration due to microbial activity, leading to changes in the biochemical, physical, and mechanical properties of treated soils [9]. Therefore, biopolymer-treated soils always exhibit unsatisfactory durability on strength.

To address the above biodegradation issue, in this study, another bio-based green material was introduced to reinforce the soil combined with biopolymer. This is because biochar's porous morphology and hydrophilic nature may have the potential to inhibit the rapid hydrolysis of biopolymers and slow down the process of biodegradation [10–12]. A series of static and dynamic triaxial tests was conducted to investigate the reinforcement effects. To quantify the biodegradation, two indicators (water-soluble carbohydrates and β-glucosidase activity) and soil's shear strength parameters (internal friction angle and cohesion) were measured after curing samples for 1, 7, 14, 21, and 28 days. Scanning electron microscopy (SEM) test was conducted to illustrate the change of microstructures due to biodegradation and reveal biochar's function to mitigate the attenuation of soil's shear strength.

2 Materials and Methods

2.1 Soil

The soil tested in this study was obtained from a local construction site near the UK's new high-speed railway line (HS2) in West Midland. The base soil has a specific gravity of 2.32. The liquid limit, plastic limit, and the plastic index are 24.3%, 13.9%,

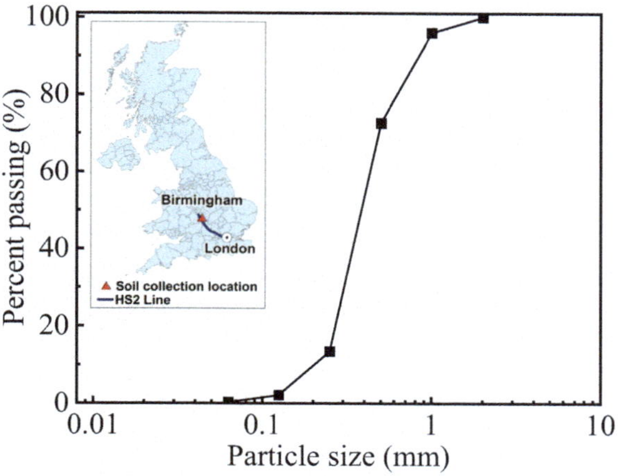

Fig. 1 Particle size distribution curve of tested soil

and 10.4. The organic matter and organic carbon are insignificant, with a proportion of 5.0% and 2.9%, respectively. The soil is slightly alkaline with a natural pH value of 7.8. Before the tests, the soil underwent double sieving through a 2-mm mesh sieve to remove large stones and remnants of vegetation. Figure 1 shows the particle size distribution curve. The soil shows low plasticity with a sand content of 89%. According to the Unified Soil Classification System (USCS), the soil is loamy sand.

2.2 Biopolymer and Biochar

Xanthan gum (XG) was selected as a biopolymer amendment. XG is an anionic polysaccharide secreted by Xanthomonas campestris bacteria. The main chain of xanthan gum is composed of cellobiose duplicate units and connected with the trisaccharide side chain by α-1,3 linkages containing d-mannose with the conjunction of β-(1,4)-d-glucuronic acid-β-(1,2)-d-mannose. This unique structure is responsible for its high viscosities in dilute solutions, outstanding thermal stability, resistance against pH changes, and good solubility [13].

The biochar used in this study was produced through slow pyrolysis for 5 h at a temperature between 350 and 500 °C. The feedstock consisted of various European hardwood arboricultural arisings, such as oak, larch, Robinia, and sweet chestnut. The biochar particles were sieved twice (2-mm sieve) to eliminate any extra-large particles and wood residuals that were not fully pyrolyzed.

2.3 Sample Preparation

Three groups of samples, untreated soil (U), xanthan gum-treated soil (X), and xanthan gum–biochar combined treatment (BX) soil, were prepared for the biodegradation test, static and dynamic triaxial tests. For the group of U, distilled water was directly added to the air-dried soil. To prepare the samples of X, 0.5% xanthan gum was added to the soil. The BX mixture was prepared by adding 0.5% xanthan gum and 10% biochar to the air-dried soil. The proportion of added materials (biopolymer and biochar) is determined by the ratio to the mass of dry soil, which was optimized in our previous research regarding the soil's shear strength improvement. Subsequently, distilled water was added to the dry soil mixture until reaching the optimal moisture content. The wet soil mixtures were filled into a cylindrical mold (diameter: 38 mm; height: 76 mm). The samples were compacted in three layers with thirty hits by a standardized Proctor hammer (275 g). All samples were saturated by the vacuum saturation method.

2.4 Triaxial Tests and Loading Conditions

The static and dynamic triaxial tests were conducted using GDS cyclic triaxial loading equipment. After saturation, the soil sample was put onto the bottom plate of the triaxial cell within a rubber membrane to isolate it from the pressurized water. The preliminary confining pressure is 20 kPa by controlling the cell pressure and back pressure remaining 320 kPa and 300 kPa, respectively, for 60 min. A B-check (pore-pressure coefficient) ratio of 0.95 was performed to ensure the sample was close to fully saturated. To simulate the in situ condition, 50 kPa was used to denote the stress condition at a depth of 4 m below track structures.

The consolidated drained static triaxial tests (CD) were performed to obtain the near in situ soil shear strength property compared to direct shear tests without considering the confining pressure and to determine the maximum deviator stresses for the calculation of CSR (cyclic stress ratio) in the cyclic triaxial tests. The confining pressure was 50 kPa, 75 kPa, and 100 kPa. The applied strain rate was 0.005%/min.

Drained dynamic triaxial tests were performed to examine samples' dynamic resilient modulus under cyclic traffic loading. A confining pressure of 50 kPa was applied to all samples under dynamic triaxial tests. The frequency of the train can be expressed as:

$$f = \frac{v}{L} \tag{1}$$

Here, f is the frequency, v (m/s) is the train's speed, and L is the distance between the bogies on one train carriage. A frequency of 2 Hz was chosen for all dynamic triaxial tests to simulate the speed of a Pendolino train ($L = 25$ m), which is an ordinary train type in the United Kingdom. The simulated train speed was 200 km/h. Three cyclic stress ratios (CSRs), 0.4, 0.6, and 0.8, were applied to the soil samples. All soil samples were subjected to 100,000 dynamic loading cycles.

2.5 *Quantification of Biodegradation*

As shown in Fig. 2, biodegradation in soil encompasses complex biochemical processes. To quantify the degradation of xanthan gum over time, the concentration of water-soluble carbohydrates (WSC) was selected as an indicator of residual xanthan gum, and β-glucosidase activity was chosen as a biomarker for the decomposition products. In this study, the concentration of WSC and β-glucosidase were measured by total carbohydrates test kit (Merck Life Science UK Limited) and β-glucosidase test kit (Merck Life Science UK Limited), respectively. Biochemical tests were performed following the curing of samples for 1, 14, 21, and 28 days. Consequently, the internal friction angle and cohesion were measured to determine the soil's static strength. Scanning electron microscopy (SEM) was employed to observe the microstructural characteristics of the samples to elucidate the mechanisms behind biochar's effect on xanthan gum degradation and changes in soil strength.

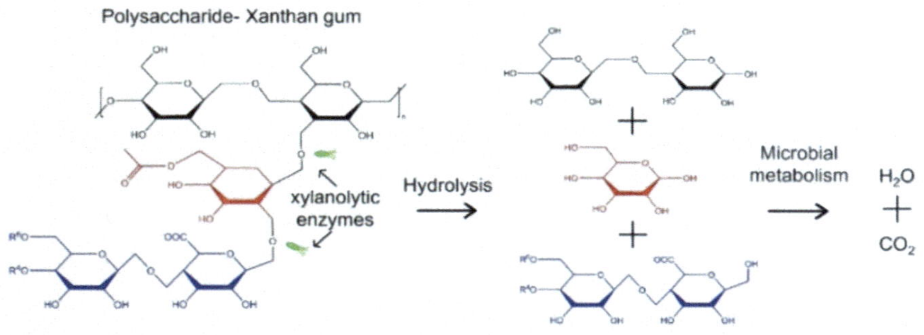

Fig. 2 Biochemical processes of xanthan gum's degradation

3 Results

3.1 Improvement of Static and Dynamic Properties

Soil Shear Strength Properties Under Triaxial Test. Figure 3 presents the failure envelope of the treated soil samples under different confining pressures. The soil was regarded as cohesionless, with a low cohesion of 2.6 kPa. It can be noted that the friction angle decreased slightly with adding XG, whereas the cohesion increased accordingly. A similar trend can be observed in other studies [14, 15]. This is believed to be the viscous nature of the XG hydrogel that enhanced the cohesion of the treated soil. However, the friction angle decreased accordingly due to the hydrogel's separation of the sand particles. A similar trend can be observed for biochar-treated soils with a neglectable change in the internal friction angle and a slightly higher improvement in cohesion. This increase in the cohesion force can be attributed to the abundant surface functional groups of the biochar, which are affected by the cementation due to solute precipitation [16]. The improvement in soil cohesion under saturated conditions has also been reported by [17, 18]. On the contrary, the soil under the combined treatment shows a significant reduction (53%) in the internal friction angle and increased cohesion from 2.6 kPa to 52 kPa, showing a dramatic improvement in the cementation of the soil particles. This enhanced aggregation effect is believed to be due to the polymerization between the XG hydrogel, biochar particles, and soil particles.

Effect of CSR on the Resilient Strain and Resilient Modulus. The evolution of the resilient strain and resilient modulus (M_r) during cyclic loading is presented in Fig. 4. Similar to the accumulative plastic strain, it can be observed that the resilient strain showed a gradually decreasing trend with the increase of loading cycles. At 0.4

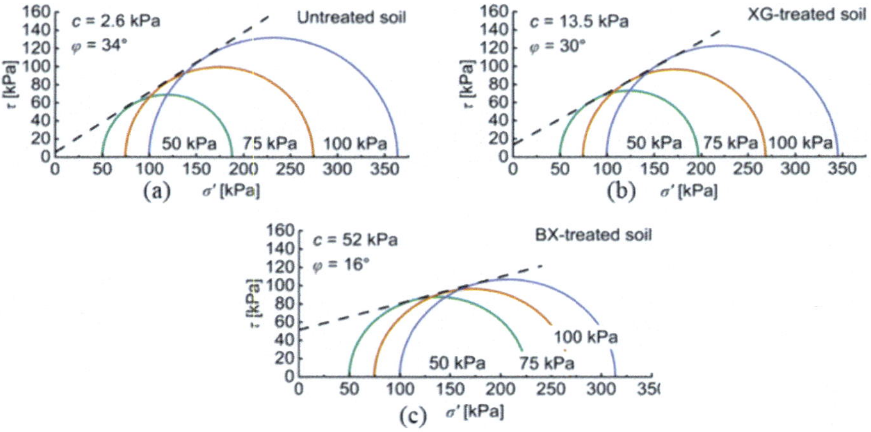

Fig. 3 Mohr circle failure envelope for treated soils under consolidated drained triaxial test

CSR, the resilient strain and resilient modulus decreased and increased, respectively, indicating that the excess pore pressure (EPP) dissipated smoothly. The resilient strains for untreated, XG-treated, and BX-treated are shown in ascending order, indicating that the treatment had limited effects on the resilient modulus of the base soil. However, at 0.6 CSR, a slight increase in both resilient strain and modulus was noted for XG-treated soils, indicating an improvement in the soil stiffness at the first 100 cycles (Fig. 4b). After that, the increasing rate of the resilient modulus gradually decreased and became similar to that of untreated soil toward the end of the test. A noticeable turning point can be observed for the BX-treated soil sample, indicating a softening process for the soil before the fact. This change in curves can be explained by the accumulation of EPP at the early stage of the test due to the low k_{sat} of the BX-treated soils. This is attributed to the enhanced aggregation effect and the polymerization among particles. After the dissipation of EPP, the resilient modulus gradually increased. Similar changes in the curves can be observed for BX-treated soil samples at 0.8 CSR due to the sudden increases of the cyclic stress and the accumulative EPP, as shown in Fig. 4c. Many fluctuations were noted in the untreated soil sample. This phenomenon might be attributed to the cohesionless nature of the base soil, which means that multiple failure planes and cracks might be initiated due to the high cyclic stress and limited cohesive force within the base soil.

The dynamic properties of X-treated soil are close to those of the untreated soil. However, after combining with biochar, a significant reduction in the dynamic resilient modulus was investigated. This reduction becomes more pronounced with an increasing number of loading cycles. With the CSR of 0.4, 0.6, and 0.8, the maximum decreases in the resilient modulus are 57.33%, 59.65%, and 63.37%, respectively.

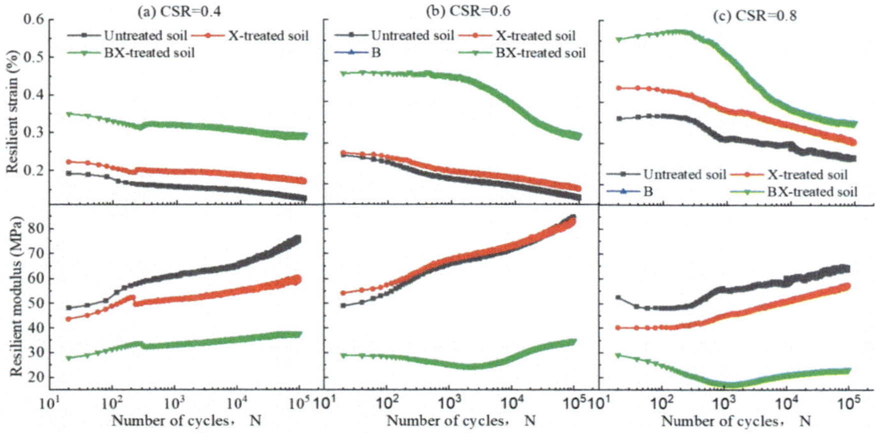

Fig. 4 Resilient strain and resilient modulus of treated soils under **a** 0.4 CSR, **b** 0.6 CSR, and **c** 0.8 CSR

3.2 Biodegradation and Mechanical Parameters

Figure 5a illustrates the water-soluble carbohydrates (W-S Carb) in reinforced soils over 28 days. During the first 21 days of curing, X-treated soil exhibited significantly higher levels of W-S Carb. Similarly, XG-treated soil demonstrated elevated levels of W-S Carb within the initial 14 days. In contrast, the curing time increases to 28 days, and the content of W-S Carb in untreated soil only slightly increases due to the carbohydrate content in untreated soil only dissolving slowly over time. Therefore, xanthan gum in the soil dissolves notably in the initial stage of curing, leading to the formation of hydrogels.

It is interesting that on the 21st day, the content of W-S Carb in X-treated soil decreased, dropping significantly by the 28th day with only 152 μg/g, which was even lower than the levels in untreated soil. This can be attributed to the significant hydration process of xanthan gum during the initial curing phase to form hydrogels. In the meantime, these hydrogels are decomposed by microbes, further releasing free water. This newly generated water then promotes further hydration of the xanthan gum. Accompanied by an increasing content of hydrogels, the microbial population in the soil gradually proliferates, thus accelerating the degradation of the hydrogels. In untreated soil, the microbial community remains relatively stable, explaining by the 28th day, the W-S Carb in X-treated soil were lower than that in untreated soil.

For XG-treated soil, the water-soluble carbohydrates (W-S Carb) formed during the initial curing phase were decomposed by soil microbes, leading to a significant reduction starting from the 14th day. In the later stages of curing, the hydrolysis of xanthan gum continued at a slow pace, and W-S Carb showed a gradual increasing trend. This means that in the early stages, XG-treated soil behaved similarly to X-treated soil, but in the later stages, it resembled untreated soil. This is because biochar influences the movement of free water within the soil, resulting in the slow release of water during the later stages of curing. Consequently, the addition of biochar effectively inhibited the hydrolysis of xanthan gum.

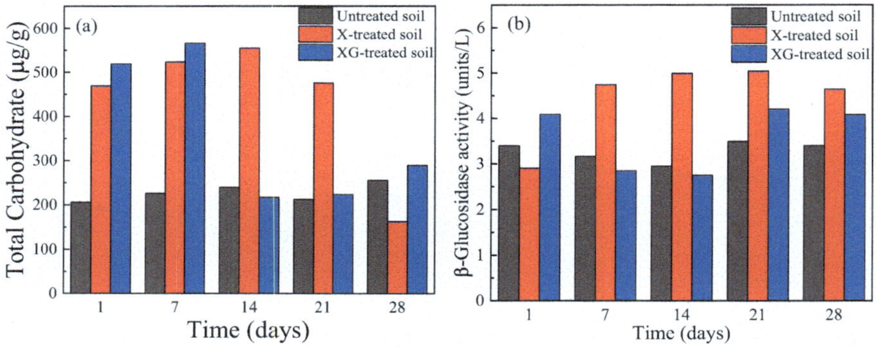

Fig. 5 The change of total carbohydrate (**a**) and β-glucosidase activity (**b**) of reinforced soil

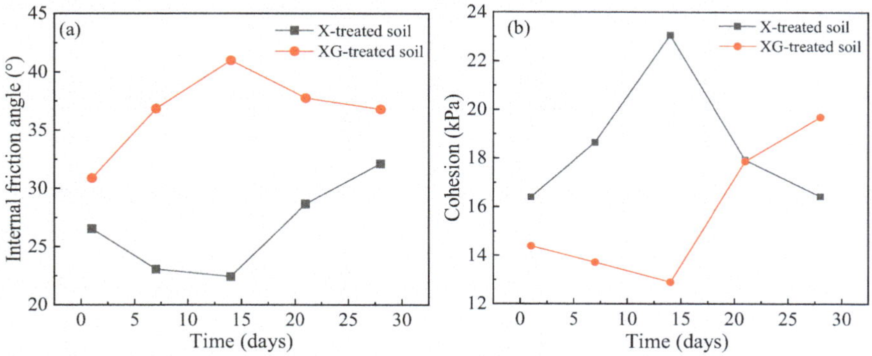

Fig. 6 The change of internal friction angle (**a**) and cohesion (**b**) over time

Figure 5b presents the contents of β-glucosidase activity at different curing days, which directly reflects the extent of biodegradation. In untreated soil, β-glucosidase activity fluctuated slightly over time, indicating a stable rate of hydrogel degradation. For X-treated soil, β-glucosidase activity peaked on the 14th day and subsequently decreased. This reduction is attributed to the decrease in xanthan gum content and the corresponding reduction in hydrogel content. However, with the addition of biochar, the changes in β-glucosidase activity in XG-treated soil exhibit a contrary pattern. The effect of biochar on the xanthan gum's hydrolysis results in limited hydrogels in the soil. As biochar slowly releases free water, the hydrogel content gradually increases, making the degradation reactions more significant. Consequently, β-glucosidase activity reaches its lowest level on the 14th day and then gradually increases.

Figure 6 shows the change of internal friction angle and cohesion with curing time. It is obvious that the peak values occur on the 14th day, indicating that the biodegradation process significantly influences the soil's mechanical properties. After 28 days of curing, the internal friction angle and cohesion of XG-treated soil increased by 19.35% and 34.49%, respectively. For X-treated soil, the internal friction angle increased by 18.51%. The cohesion of X-treated soil significantly increased due to produced hydrogels but decreased back to its initial level due to biodegradation. It can be anticipated that, with the continuation of the decomposition process, the cohesion of X-treated soil will continue to decrease. Therefore, the combining of xanthan gum and biochar can offer better durability on soil reinforcement.

3.3 *Scanning Electron Microscopy (SEM)*

Figure 7a and c show the SEM images of X-treated and XG-treated soil, respectively, after the preparation of samples. In X-treated samples, clear hydrogel bonds among soil particles were investigated, which contributes to the increased cohesion of the soil. Similarly, in XG-treated soil, a fibrous structural film formed by hydrogels can be observed. Hydrogels adhered to the surface of biochar and aggregated in biochar's pores. Additionally, due to the angular morphology characteristics of biochar, its particles are more prone to interlocking and embedding. Under external forces, this microstructure prevents the particles from sliding relative to each other, thereby exhibiting a higher internal friction angle. As shown in Fig. 7b, after 28 days of curing, sparse hydrogel bonds can still be observed on the surface of soil particles, although they have partially fractured. The reduction in these inter-particle bonds has led to a decline in the soil's cohesion. The degradation of hydrogels increased the friction between particles, leading to an increase in the internal friction angle. Figure 7d presents the microstructure of BX-treated soil on the 28th day. Due to the abundant fine pores at the biochar's surface and hydrophilic functional groups [11], the free water within the soil was stored in the biochar, thereby slowing down the hydrolysis process of xanthan gum. As the hydrogel produced by hydrolysis gradually increases, the overall strength of the soil shows a trend of gradual growth. However, after 14 days of curing, the hydrogels acted as a "lubricant" between the particles, reducing the internal friction angle.

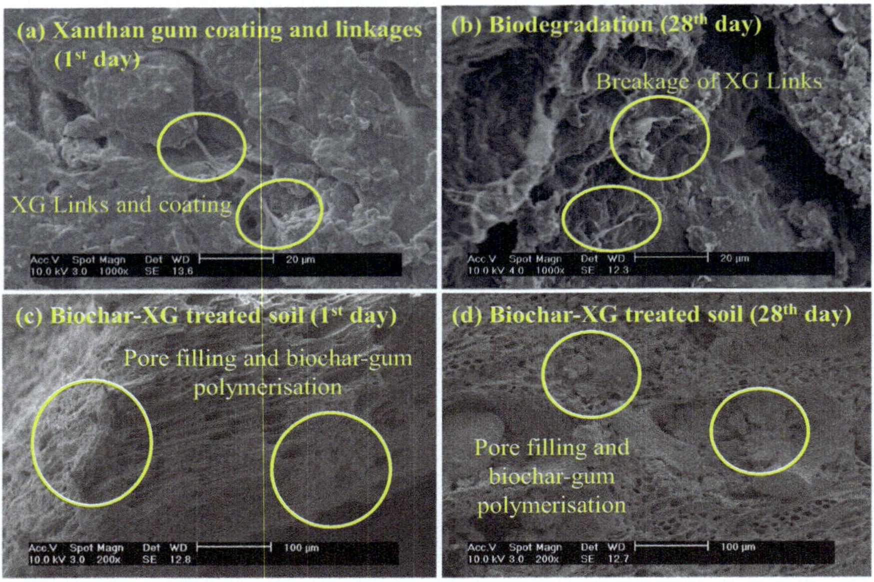

Fig. 7 SEM images of **a** X-treated soil on the 1st day, **b** X-treated soil on the 28th day, **c** XG-treated soil on the 1st day, and **d** XG-treated soil on the 28th day

4 Conclusions

This study investigated the biodegradation of xanthan gum in stabilized soil and the mitigation method by combining it with biochar. It was investigated that xanthan gum forms hydrogels through hydrolysis reactions and establishes hydrogel bonds among soil particles, enhancing the soil's static and dynamic strength. However, microbial can degrade these hydrogels, resulting in a significant attenuation of soil reinforcement, especially after 14 days of curing. Due to biochar's porous morphology and hydrophilic functional groups, the free water in the soil is gradually released and participates in the hydrolysis of xanthan gum. Additionally, hydrogels are captured by the voids in the biochar. This results in stronger inter-particle binding. Consequently, the strength of XG-treated soil exhibits a trend of slow growth over time. Biochar's angular structure enhances interlock and embedment among particles, leading to a 42% increase in shear strength compared to untreated soil, and a 21% increase compared to X-treated soil. Therefore, it is advisable to use the combined approach of xanthan gum and biochar to reinforce soil in the engineering practice.

References

1. Kendon M, McCarthy M, Jevrejeva S, Matthews A, Sparks T, Garforth J (2021) State of the UK climate 2020. Int J Climatol 41:1–76.20
2. Ni J, Geng X (2022) Radial consolidation of prefabricated vertical drain-reinforced soft clays under cyclic loading. Transp Geotech 37:100840
3. DeJong JT, Mortensen BM, Martinez BC, Nelson DC (2010) Bio-mediated soil improvement. Ecol Eng 36(2):197–210
4. Cheng Z, Geng X (2023) Mechanical behaviours of biopolymers reinforced natural soil. Struct Eng Mech 88(2):179–188
5. Ni J, Li S, Geng X (2022) Mechanical and biodeterioration behaviours of a clayey soil strengthened with combined carrageenan and casein. Acta Geotech 17(12):5411–5427
6. Cheng Z, Geng X (2023) Investigation of unconfined compressive strength for biopolymer treated clay. Constr Build Mater 385:131458
7. Chen C, Wu L, Perdjon M, Huang X, Peng Y (2019) The drying effect on xanthan gum biopolymer treated sandy soil shear strength. Constr Build Mater 197:271–279
8. Ni J, Geng X (2023) Vegetation growth promotion and overall strength improvement using biopolymers in vegetated soils. Can Geotech J Just-IN
9. Vydehi KV, Moghal AAB (2022) Effect of biopolymeric stabilization on the strength and compressibility characteristics of cohesive soil. J Mater Civ Eng 34(2):04021428
10. Rizwan M, Ali S, Qayyum MF, Ibrahim M, Zia-ur-Rehman M, Abbas T, Ok YS (2016) Mechanisms of biochar-mediated alleviation of toxicity of trace elements in plants: a critical review. Environ Sci Pollut Res 23:2230–2248
11. Pardo GS, Sarmah AK, Orense RP (2019) Mechanism of improvement of biochar on shear strength and liquefaction resistance of sand. Géotechnique 69(6):471–480
12. Zong Y, Xiao Q, Lu S (2016) Acidity, water retention, and mechanical physical quality of a strongly acidic Ultisol amended with biochars derived from different feedstocks. J Soils Sediments 16:177–190
13. Nasrollahzadeh M (2021) Biopolymer-based metal nanoparticle chemistry for sustainable applications: volume 2: applications. Elsevier, p 15

14. Lee M, Im J, Cho GC, Ryu HH, Chang I (2020) Interfacial shearing behavior along xanthan gum biopolymer-treated sand and solid interfaces and its meaning in geotechnical engineering aspects. Appl Sci 11(1):139
15. Smitha S, Rangaswamy K, Keerthi DS (2021) Triaxial test behaviour of silty sands treated with agar biopolymer. Int J Geotech Eng 15(4):484–495
16. Lu Y, Gu K, Shen Z, Tang CS, Shi B, Zhou Q (2023) Biochar implications for the engineering properties of soils: a review. Sci Total Environ 164185
17. Kumar H, Huang S, Mei G, Garg A (2021) Influence of feedstock type and particle size on efficiency of biochar in improving tensile crack resistance and shear strength in lean clayey soil. Int J Damage Mech 30(4):646–661
18. Ng CWW, Cai W, So PS, Liao J, Lau SY (2022) Effects of biochar on shear strength of completely decomposed granite. Environ Sci Pollut Res 29(32):49422–49428

The manufacturer's authorised representative in the EU is Springer Nature Customer Service Centre GmbH, Europaplatz 3, 69115 Heidelberg, Germany. If you have any concerns regarding our products, please contact ProductSafety@springernature.com

Printed and bound by CPI Group (UK) Ltd, Croydon, CR0 4YY
01/10/2026
02225962-0003